PRISON MEDIA

Distribution Matters

Edited by Joshua Braun and Ramon Lobato

The Distribution Matters series publishes original scholarship on the social impact of media distribution networks. Drawing widely from the fields of communication history, cultural studies, and media industry studies, books in this series ask questions about how and why distribution matters to civic life and popular culture.

Emily West, *Buy Now: How Amazon Branded Convenience and Normalized Monopoly*

Robin Steedman, *Creative Hustling: Women Making and Distributing Films from Nairobi*

Anne Kaun and Fredrik Stiernstedt, *Prison Media: Incarceration and the Infrastructures of Work and Technology*

Lee McGuigan, *Selling the American People: Advertising, Optimization, and the Origins of Adtech*

PRISON MEDIA

INCARCERATION AND THE INFRASTRUCTURES OF WORK AND TECHNOLOGY

ANNE KAUN AND FREDRIK STIERNSTEDT

The MIT Press
Cambridge, Massachusetts
London, England

The MIT Press would like to thank the anonymous peer reviewers who provided comments on drafts of this book. The generous work of academic experts is essential for establishing the authority and quality of our publications. We acknowledge with gratitude the contributions of these otherwise uncredited readers.

This book was set in Bembo Book MT PRO by Westchester Publishing Services. Printed and bound in the United States of America.

Library of Congress Cataloging-in-Publication Data

Names: Kaun, Anne, author. | Stiernstedt, Fredrik, author.
Title: Prison media : incarceration and the infrastructures of work and technology / Anne Kaun and Fredrik Stiernstedt.
Description: Cambridge, Massachusetts : The MIT Press, 2023. | Series: Distribution matters | Includes bibliographical references and index.
Identifiers: LCCN 2022036840 (print) | LCCN 2022036841 (ebook) | ISBN 9780262545495 (paperback) | ISBN 9780262374330 (epub) | ISBN 9780262374347 (pdf)
Subjects: LCSH: Prisons. | Power (Social sciences) | Authority. | Mass media.
Classification: LCC HV8665 .K38 2022 (print) | LCC HV8665 (ebook) | DDC 365—dc23/eng/20220822
LC record available at https://lccn.loc.gov/2022036840
LC ebook record available at https://lccn.loc.gov/2022036841

10 9 8 7 6 5 4 3 2 1

To all the Kennet Ahls in the world

To all the Kenneth Aldis in the world?

CONTENTS

PREFACE

In 2019 there were 4,834 incarcerated individuals in Swedish prisons; 94 percent of them were men, and 6 percent were women. The majority were convicted of violent crimes in connection with drug offenses and smuggling. Worldwide there were more than 10.35 million incarcerated individuals in pretrial institutions, remand prisons, and low-, medium-, and high-security prisons, a number equaling the total population of Sweden. And while the numbers of incarcerated persons are expected to be growing over the course of the next decade and the Swedish Prison and Probation Service is preparing for yet another large-scale construction boom extending existing facilities and building completely new ones, there is comparatively little discussion about the conditions of incarceration in Swedish prisons. Only rarely does the public put its eyes on Swedish prisons and remand prisons, as in the case of the artist ASAP Rocky, who was arrested in 2019 for aggravated assault and was held in Kronoberg's jail for more than a month,[1] an episode that has now been turned into the documentary *Stockholm Syndrome,* and when the mounting critique of the conditions in Swedish remand prisons is issued over and over again by different committees including the returning report from the Council of Europe's Committee for the Prevention of Torture and Inhuman or Degrading Treatment or Punishment (Council of Europe, 2021). This and other reports criticize the extensive periods of isolation at Swedish remand prisons, which are considered a violation of human rights. It is in these moments of scrutiny that prisons and the role of punishment gain some visibility in public discourse. However, to a large extent the public remains silent and indifferent. Working on this book has made this silence and disinterest very palatable, while it also made obvious how closely prison worlds and our social lives are interwoven. This book is an attempt—an intervention—to make these connections visible.

We were lucky enough to be supported in this endeavor. The research for this book was supported by smaller grants among others from the platform for police-related research at Södertörn University. Tina Askanius, Alessandro Delfanti, Staffan Ericson, Karin Fast, Ingrid Forsler, Max Kunze, Aaron Shapiro, and Patrik Åker have read and commented on parts of the manuscript. Thanks also to our colleagues at the Department for Media and Communication Studies at Södertörn University who endured years of presentations on prison-related stuff. We are done; finally, we are done. We are also grateful for the opportunities to present the full manuscript during research seminars across Sweden including at Jönköping University, Stockholm University, Uppsala University, Karlstad University, and Umeå University. The discussions and questions were truly inspiring and helped shape the arguments of the book. We are also grateful for the support and openness of the staff at different Swedish high-security prisons who showed us around and shared their experiences with us. The insights gained there were instrumental for large parts of the book, and so has been the support by archivists at the Swedish National Archive who helped us navigate the vast collections of the Swedish Prison and Probation Service. We also would like to extend our gratitude to the ladies from the Red Cross Remand Prison Visitor Groups. They are doing such important work, they are an inspiration in human kindness and generosity, and they gave us hope for the future. Parts of the book were researched and written at the Sigtuna Foundation, academic heaven on earth, where we spent separate research stays. We are grateful to the Ekman Foundation for opening this world of intellectual and spiritual freedom to us.

We dedicate this book to all those who have experienced incarceration and all those who try to make incarceration more bearable and imagine alternative ways.

1

INTRODUCTION: WHY PRISON MEDIA MATTER

In 1974 the formerly incarcerated Lasse Strömstedt and Christer Dahl, with personal experience from one of the largest prisons in Sweden, Hall Prison, published the novel *Grundbulten* (The foundation bolt) under the pseudonym Kennet Ahl. The story of the crime novel unfolds at the prison, which is depicted as a "small world of concrete," a "beloved hell" in the countryside south of Stockholm. The title is adapted from one of the main characters in the novel who is obsessed with finding the "foundation bolt" of the prison, the one piece that holds the whole construction together. In a desperate attempt to find this one magic screw, this man walks around with a stolen screwdriver and dismantles every screw he can find in the prison until he is discovered and punished. The narrator explains that "a foundation bolt, as you might understand, is the bolt that holds the foundation of something, such as a prison. If you find the foundation bolt and remove it, the prison collapses" (Ahl, 1974, p. 9). The foundation bolt serves as metaphor for the search of the social foundation, the one thing that holds society together in its entirety, and if dismantled, total destruction awaits. The backdrop for this existential story is the prison, a place with its own rules, norms, and language, a place of contradictions ("beloved hell") and of limited scope ("small world of concrete") (Ahl, 1974, p. 9). In the novel, the prison is presented as a medium in a metaphorical sense, standing in for the society as a whole, and in a very literal sense as a medium for circulating information, channeling communication, connecting, and disconnecting people. Especially the culvert system and closed-circuit television (CCTV) surveillance are a returning motif in the story, structuring the movement and communication of the incarcerated individuals at Hall as they are presented as central mechanisms of dehumanization.

The major trope of *Prison Media: Incarceration and the Infrastructures of Work and Technology*, our foundational bolt, is prison media, a term that captures both media that are produced in and for the prison and the prison as a medium. Media are here understood in terms of their material properties, as infrastructures and artifacts as well as practices rather than content or forms of representation (Williams, 1974; Couldry, 2012). We explore the double nature of prison media: the entanglement of media infrastructures inside and outside the prison through *prison work*, how *prison architecture* serves as a medium in itself, and in what ways prisons constitute test beds for *technology development*. Our main starting point for engaging with prison media is that the modern penal system is deeply intertwined with both media and communication. In recent years, this connection has taken new forms with the development and implementation of new digital media technologies in the prison context. For example, Hong Kong plans to introduce so-called smart digital technologies for the surveillance and control of its 8,300 incarcerated individuals. The aim is to make correction facilities safer and more efficient. With the help of smart devices, including robots that scan for contraband as well as personal monitoring devises that measure the pulse and other bodily functions of incarcerated individuals, the guards and wardens are able to preempt violations of prison rules as well as self-harm, according to Hong Kong's Correctional Services Department. Besides robots searching for contraband and wristbands monitoring the bodies of incarcerated people, the prisons are to be equipped with artificial intelligence–enabled cameras that detect suspicious behavior and unusual movement patterns among the prison population.[1] Similarly, Singaporean prisons have implemented the so-called prisons without guards program. The aim is to free guards from mere surveillance and control tasks with the help of smart technologies such as radio-frequency identification wristbands to automate counts of the incarcerated individuals and digital kiosks that allow incarcerated individuals to order specific products, schedule family visits, and follow up on their requests. The extensive deployment of smart CCTV cameras that are using machine learning is part of the program. In promotional articles and videos, the Ministry of Home Affairs promises that with the help of smart technologies, guards will be turned into personal supervisors for incarcerated individuals with a focus on psychological support rather than control.[2]

These examples illustrate that at the moment, digital technologies are imagined as changing the ways in which incarcerated people *are doing their time*. The imaginaries and promises range from better surveillance and more efficiency to a strengthened focus on support and rehabilitation through digital technologies and by freeing up time for the staff so they can prioritize support in regard to control of the incarcerated individuals. But smart technologies and their contributions to the organization of prisons are only one contemporary way in which media matter for the corrections sector and corrections matter for media.

The intertwining of prisons and media includes a broad range of technologies, practices, and forms of communication and, as we will show throughout the book, is in no way a novel phenomenon that emerges first in relation to smart digital technologies. On the contrary, prisons and media have been closely related since the emergence of the modern prison in the nineteenth century. Initially banning media to gradually allowing specific kinds of media as part of the corrective approach, the penal system has always considered media to both foster certain corrective approaches and undermine prison rules, as in the case of contraband. Already in the modern cell prisons of the nineteenth century and even more so in the industrial prisons of the twentieth century, incarcerated individuals in very concrete ways have been linked to the development of media technologies: incarcerated individuals through their work have produced important parts of media infrastructures and technologies for communication and mediation. Prison architecture has furthermore historically and up to today been an answer to the question of communication: the buildings themselves allow and constrain specific forms of communication and hence function as an example of how the built environment constitutes an important trajectory of mediation. Media technologies—from early experiments of photography in the nineteenth century and onward—have been and are tested in prisons to later transition into other sectors and areas of everyday use. Prisons have hence in many ways been test beds for technological development, with repercussions for the broader society. At this moment when technology is imagined as transforming prisons, it makes good sense to return to the longer history of the relation between prisons and media to gain a better understanding of their contemporary and future configurations.

Based on these starting points, the book engages with prisons as crucial for our modern media infrastructures and provides a historical analysis of how our media infrastructures have been built and maintained by incarcerated individuals and imagined and developed in and through the prison. In that sense, we follow the French anthropologist Didier Fassin's (2017) approach to writing a media theory through the lens of the prison system. To begin constructing such a theory of prison media, we must first understand what kind of space prisons are.

Michel Foucault not only traced the emergence of the modern prison regime (Foucault, 1975) but also suggested prisons as a prime example of heterotopian places defined as concrete places that exist and emerge in the founding of society (Foucault, 1984). Because heterotopian places are still located in reality, they are not nonplaces. In more idealist terms, they emerge as dark mirrors of utopian imaginations while being distinct from dystopia. Although they are in that sense fundamental to society, they also present countersites that invert the norms corresponding to other sites within society, simultaneously representing and contesting them. Although prisons and prison systems have changed over time and vary in the details of their structure and methods of control, they fulfill a specific social function across modern cultures, namely, to contain individuals who deviate from what has been defined as the normal and required way of being in society. The modern prison has from the outset been about not only detaining deviant criminals but also first and foremost reforming those criminals. How to achieve this rehabilitation has, however, shifted over time.

Prisons also purport to reconcile in one single place several sites that might otherwise be incompatible with each other: they are at the same time both institutions of education, work, and leisure and temporary homes that collapse experiences and relationships otherwise linked to separate spheres in life. Prisons are integral to society and to capitalist economic systems by stabilizing and maintaining deviant populations at the same time that their lack of public visibility and public access relegates them to the margins of society. Prisons also establish a complete break with traditional understandings and experiences of time; they form heterochronies captured in idiomatic expressions such as "doing time" when time is experienced as both being paused and fast-forwarded. Prisons are heterotopias in the sense of being removed from the public sphere. Members of the general public do not generally have access to them without special permission. Finally, prisons establish a place where all that runs counter to the good society is contained, thus making the good society possible in the first place. In that sense prisons "have a function in relation to all the space that remains" (Foucault, 1984, p. 8). In his work *Discipline and Punish*, Foucault develops a genealogy of the emergence of the modern prison as a place of punishment and also a history of discipline that is exercised upon the body through different techniques. These techniques include drills as well as timetables and exercise to coerce the body into specific behaviors. This disciplinary power is exerted through observation, judgment, and examination (Foucault, 1979). We argue that the emergence of disciplinary power through observation, judgment, and examination within and outside the prison media infrastructures is essential. In many ways these methods manifest and support disciplinary power. One site where disciplinary power of the prison emerges but that has so far been largely overlooked are media infrastructures.

INFRASTRUCTURES AND SOCIOTECHNICAL IMAGINARIES OF PRISON MEDIA

While disciplinary practices in and outside the prison have been repeatedly highlighted, the necessary media infrastructures for those practices are still being overlooked. Media infrastructures are the basic technologies, facilities, and services that make communication and information exchange possible. We explore the double role of media infrastructures, the role they play for prisons and punishment but also which roles prisons and incarcerated individuals play for our media infrastructures. Hence, we diverge in many ways from existing research of the prison-media nexus that mainly consists of accounts of the representation of prisons in popular culture and the question of media use by incarcerated individuals (Carrabine, 2008; Jewkes, 2002; Vandebosch, 2001). In this endeavor, we are relying on previous studies that have highlighted the material aspects and environmental implications of media infrastructures (Parks & Starosielski, 2015). To date, however, they have overlooked prisons as sites of media work and technology development, and the question of how this work has been an integral part of media infrastructure building and maintenance. Although media infrastructure studies have a longer history, the field has recently enjoyed a more vivid engagement, a process that John Durham Peters (2015) has described as an "infrastructural turn" in media and communication studies. Previous research on media infrastructures, according to Lisa Parks and Nicole Starosielski (2015), included, for example, studies of the imaginaries of electricians of the early power grid (Marvin, 1988), the role of telegraph messenger boys for extending communication through time and space (Downey, 2002), and the role of radio networks for extending the way of life of colonial settlers into Africa (Larkin, 2008). In recent years, a range of scholarship has contributed further to the exploration of media infrastructures by studying the role of satellite systems for global media events (Lundgren & Evans, 2017; Parks, 2005) and how undersea cables (Starosielski, 2015) and data centers (Hogan, 2015; Jakobsson & Stiernstedt, 2010, 2012; Velkova, 2016) are crucial for sustaining the internet and digital culture.

Parks and Starosielski (2015) identify three major paths of media infrastructure studies. The first path focuses on questions of work, repair, and maintenance; the second focuses on natural resources and the environmental impact of media infrastructures; and the third focuses on the question of how affect is intertwined with media infrastructures. Developing these three paths, they link their work to established traditions of media studies that combine political economy, posthumanism, and new materialism. We follow Parks and Starosielski in linking the exploration of prison media to these broader theoretical perspectives and engage with prison work, prison architecture, and prison technologies from a media and communication studies perspective.

Infrastructures are commonly understood as "that which runs 'underneath' actual structures—railroad tracks, city plumbing and sewage, electricity, roads and highways, cable wires that connect to the broadcast grid and bring pictures to our TVs. It is that upon which something else rides, or works, a platform of sorts" (Star & Bowker, 2002, p. 151). Beyond such material manifestations, infrastructures also include intellectual and institutional structures such as standards, protocols, conventions, and classification systems as well as bureaucratic forms. These are intellectual and administrative infrastructures and are the underlying systems that make both software and hardware work (Bowker & Star, 1999; Mattern, 2016). Infrastructures are hence "interrelated social, organizational, and technical components or systems" (Bowker & Star, 1999, p. 99). Infrastructures—including media infrastructures—are an assemblage of different practices in relation to material settings that are constantly evolving and continuously reshaped (Appel et al., 2018). Theorists of infrastructures have suggested that these systems are often perceived as being ready at hand and taken for granted. Infrastructures are mainly invisible until we experience a breakdown, rupture, glitch, or failure. Furthermore, infrastructures depend on large amounts of invisible work that is rarely made palatable. They are, however, not merely invisible; rather, they share the ambivalent character of being both visible and invisible. Sometimes they take center stage in our attention as things in themselves; at other times they linger in the background, overshadowed by the thing they are moving around (Holt & Vonderau,

2015). Jennifer Holt and Patrick Vonderau (2015) employ the metaphor of lenticular prints, which change with your movement and reveal different images. It is not about infrastructures changing as such; instead, it is our perspective and views on them that change. Studies of infrastructures shift their focus between what they are moving and connecting and the connective infrastructure as such. This applies to how we define prison media. As we have asserted, prisons are media in and of themselves. Their architecture connects and moves people around while relying on media technologies to organize these processes (e.g., CCTV surveillance systems, monitor control rooms, and access to television sets for incarcerated individuals). At the same time, prisons are also a site of infrastructural work: they are the places where media infrastructures are built.

Hence, prisons embody what Holt and Vonderau (2015), drawing on Susan Leigh Star and Karen Ruhleder (1996), have called the layering and bundling of systems that are distinct from each other. According to Star and Ruhleder (1996), infrastructures are relational and emerge in specific practices. The infrastructure that supports a cook might not be the infrastructure that supports a communications officer, although both are working in the same building so they are related to each other. The understanding of infrastructure that we apply in these pages is fundamentally relational. It emerges in the practices and activities of people connected to prisons and their technical structures. We are examining media technologies, often considered as black boxes and invisible infrastructures, without necessarily exploring them in their technical totality (Bucher 2016). This also means that we understand—following among others Sebastian Kubitschko and Tim Schütz (2016) and Julia Velkova (2017)—infrastructuring as a practice that constantly renegotiates infrastructure. Infrastructures are fundamentally about social arrangements and must be learned rather than being adopted naturally. Following this understanding, prisons are infrastructures on different levels. On a macrosocial level, they are infrastructures that are supposed to maintain social order by controlling deviant populations. On a meso level, they arrange bodies in time and space with the help of a specific architecture of corridors, tunnels, and gates. On a micro level, they must be learned to be embodied by both guards and incarcerated individuals and structure their experience of time and space.

As an approach to the study of infrastructures, Geoffrey Bowker and Susan Leigh Star suggest "infrastructural inversion" (Bowker & Star, 1999; Star,

1999). Like reverse engineering, this approach foregrounds going backstage and the importance of looking at infrastructures in the making as well as moments of disruption as entry points to develop an understanding of these complex underlying systems (Gehl, 2014). Following this approach, we not only identify existing prison media and take them for granted but also take a closer look at how certain technologies emerged and in what ways prisons as a very specific environment have affected the outlook of media technological developments.

Prison media, however, beg not only a media infrastructural perspective that is focused on practices and relations. In order to develop our understanding of prison media, we also turn our attention to the sociotechnical imaginaries that in many ways undergird, prepare, and legitimize the material formation of media infrastructures. Sociotechnical imaginaries are visions that sometimes materialize and are translated into concrete practices. They are "collectively held, institutionally stabilized, and publicly performed visions of desirable futures (or of resistance against the undesirable), and they are also animated by shared understandings of forms of social life and social order attainable through and supportive of advances in science and technology" (Jasanoff, 2005, p. 19).

Such imaginaries are hence more than merely individual ideas. They are durable, collective, and performative but, at the same time, never stable and change over time. Astrid Mager and Christian Katzenbach (2021) also remind us that imaginaries are always contested and under negotiation. There is no given agreement of what the future should look like, and this contested and conflictual nature of sociotechnical imaginaries is of interest here. The articulation of what technology means and does is always also translational work between different sectors, as Michael Hockenhull and Marisa Leavitt Cohen (2021) show. They argue that corporations active in the tech sector enact sociotechnical imaginaries of the digital future by producing hot air and the aura of technological cool at intersectorial conferences and trade shows with the aim of pushing digital technologies into, for example, the public sector. Similarly, we consider the discursive construction of prison media as performative and forms of meaning making that have consequences for how prisons are organized but also for how technology is developed for sectors other than corrections. Drawing on and further developing the above discussions, we engage with prison media empirically through three interrelated perspectives, namely work, architecture, and technology development as important practices, sites, and aspects of infrastructures.

WORK

Critical infrastructure studies have turned our attention to the social practices of and the work that goes into developing, building, repairing, and maintaining infrastructures of communication (Parks & Starosielski, 2015). Susan Leigh Star (1999) has pointed to the importance of making visible the ordinary work that is involved in infrastructure building and maintenance, including the work of janitors and cleaners (see also Parks & Starosielski, 2015). This involves studying the work that emerges with and through media infrastructures such as Downey's work on telegraph messenger boys. The messenger boys constitute an example of an emerging profession at the intersection between automation and physical work, professions that combine mechanical with human labor. The integration of analyses of work and labor with questions of infrastructure hence points to their sociotechnical character and that technological infrastructures have to rely on human productive activity. Work and labor in modern capitalist society always implies power, authority, and possibility for resistance, and these dimensions are crucial for understanding why and how infrastructures are realized and how they work. The work that goes into building, maintaining, and repairing infrastructures has often been invisible and gendered and has had a low status. Within media studies more generally, it has not been understood as or included in analyses of "media work" (Deuze, 2007) or in other discussion of work in the cultural industries to any high degree. An exception to this is Vicki Mayer (2011), who has called attention to the television production workers "below the line" that are normally not included in discussions about the creative industries. Expanding the notion of television producers, Mayer includes assembly-line workers in electronic factories in Brazil, soft-core videographers, casting personnel for reality shows, and citizen volunteers to local and regional television committees (public utilities committees) and argues for their contributions to television productions that are invisible and rarely valued. Similarly, Edwards et al. (2009) have called for more attention to "those at the receiving end of infrastructure—those who are subjected to its distribution regimes and marginalizations in everyday life" (p. 372). This includes the workers and producers of infrastructure as well as the users.

We assert that marginalized media work is conducted not only below the line but also behind bars. From the telegraph and the canal network to machine learning and artificial intelligence, prisons and incarcerated individuals have been a crucial—but largely invisible—workforce underpinning media infrastructures and networks of communication. Therefore, we emphasize the idea that prisons are constructed for the sake of work (Tornklint, 1971) and that media-related work has been a crucial part of the labor conducted by incarcerated individuals. At the same time, media work conducted in prisons has contributed in decisive ways to the emergence and maintenance of several media infrastructures, including the telegraph system, the postal system, television, and radio. To explore prison media work hence opens a reassessment of how our contemporary media society has developed and engages in the question of how media culture relates to subaltern lives and experiences and how the communication infrastructures of the twentieth century have been affected by their development and maintenance "behind bars."

ARCHITECTURE

If media infrastructures emerge through practices and in relations between users and material systems, the spatial and architectural aspects of infrastructure are fundamental to understanding those relations. Efforts to foreground the physical and aesthetic character of the buildings, pipes, towers, and cables that form our media stem from a longer tradition of exploring media with and through the notion of space. Consequently, several media scholars have earlier explored the changes in spatiality in relation to media. Joshua Meyrowitz (1986), for example, argues in *No Sense of Place* that electronic media have created new social situations that are independent from locality. Similarly, John B. Thompson (1995) speaks of "despatialized simultaneity" when referring to publics that are independent of locality. Discussions of mediated communication that fundamentally engage space as a relevant category have led to a "spatial turn" in media communication studies, which has directed attention to the spatial contexts of media usage instead of arguing for the end of space (Jansson & Falkheimer, 2006). For instance, Ericson et al. (2010) argue that our homes, cities, and shopping malls became the spatial prism through which media scholars developed an understanding of communication practices that are increasingly without boundaries.

Not only have scholars explored how space structures communication, but they have also engaged with the aesthetic dimensions of infrastructures and highlight their strategic visibility. Brian Larkin (2018) argues that architectural aesthetics constitute "a form of political action that is linked, but differs from, their material operations. And political aesthetics is one way that we can understand the promise of infrastructures" (p. 176). When we explore infrastructures' architectural aesthetics, it is important to interrogate what is made visible and what is made invisible. Infrastructures are sometimes spectacular structures that dominate urban and rural spaces, and sometimes they are made to blend in and become invisible. Some cases combine both visible and invisible features. The visibility and invisibility of infrastructures are an integral part of their emergence in the context of technical, political, and representational processes. In that sense, they are actively achieved and negotiated and hence are always more than just pipes or cables supplying water, electricity, or the internet (Larkin, 2018).

If the architecture of infrastructures is fundamentally about structuring and negotiating communication, prison architecture itself could be understood as a medium that includes walls, fences, and sounds, smells, and tastes that structure time, space, and communication for the incarcerated individuals as well as guards and visitors as other media such as the television and the computer are doing (Fassin, 2017). Prison media architecture takes shape in the imaginaries of politicians, state planners, and architects who envision, plan, and implement the fundamental structure of prisons that is crucial for the communication possibilities of incarcerated individuals, guards, and visitors while also giving form to ideas about punishment and rehabilitation. Studying the architecture and the visions that inform the architectural layout related to the built environment of the prison reveals also how communication is allowed and constrained. At the same time, incarcerated individuals and visitors develop often creative ways of circumventing these constraints and appropriate the architecture to develop unexpected, daring, and defiant ways for unsanctioned communication.

DEVELOPMENT

Finally, we can consider infrastructures in terms of technological development and imaginaries mobilized during the implementation process. Charles Taylor (2003) argues that modernity relates to a set of new social imaginaries and a new moral order of society that includes the market economy, the public sphere, and the self-governing people. In order to understand modernity, then, we need to engage with the self-understandings that are constitutive of modernity. This includes the role of technologies that are entangled with self-understandings in later modern societies. Sociotechnical imaginaries offer a way to engage with the question of how we imagine our social existence in relation to others, fostered by technologies and the future visions connected with technological development. Furthermore, social imaginaries capture expectations and the normative notions that underlie these expectations. Sociotechnical imaginaries are shared by a large group of people and constitute a common understanding that allows for common practices and a shared sense of legitimacy. These imaginaries crystallize in images, stories, and legends but also institutions and the way in which institutions are organized as well as their underlying infrastructures. Specific societal institutions such as courts, traditional media, and legislative bodies have the power to highlight certain sociotechnical imaginaries over others. For example, strategy documents of the Swedish government highlighting the potential and importance of artificial intelligence to solve crucial societal challenges work to materialize sociotechnical imaginaries. These imaginaries not only relate to specific ideas about technologies but also encompass normative ideas about how the good life and society ought to be. In that sense, they express and reify a shared sense of good and evil (Jasanoff, 2015). Hence, sociotechnical imaginaries carry normative aspects while always remaining provisional, imperfect, and under construction (Willim, 2017). Beyond their aspirational character, sociotechnical imaginaries are also expressions of mundanization (Willim, 2017). Through concrete ways of imagining the workings and utilities of technologies—for example, imagining the physical distance that emails travel across space or giving algorithms names to make them more palatable—we integrate them into our everyday lives and make them graspable beyond their ungraspable complexity.

Sociotechnical imaginaries become especially important in the context of prisons intended to hold and maintain populations understood to be dangerous or undesirable. Prisons and the way they are organized become concrete and institutional expressions of social imaginaries that stabilize and perform ideas about the good society. As such, prisons are spaces that encapsulate modern social imaginaries by both expressing and constituting the distinction between the good and the bad. Furthermore, because prisons are used as a space to imagine and develop new technologies, we argue that they also function as places where social imaginaries about media crystallize and are put to the test. Ideas about which forms of media are "good" and "bad" are expressed in the regulations of media access in prisons as well as by testing technologies to control prison populations.

MEDIA DISTRIBUTION IN THE PRISON SYSTEM

Prison media work, architecture, and development constitute a complex system of practices and relations that are media infrastructures, and they have a longer history. From the outset of modern prisons in the nineteenth century, the question of how to regulate incarcerated individuals' media access and communications has been a central issue. The management of communication between incarcerated individuals, between guards and incarcerated individuals, and between incarcerated individuals and visitors from outside is crucial for the establishment and maintenance of control. Not only are there national regulations of what communications are allowed (for the Swedish context in the *Fängelseförordning*—Prison regulations from 2010), but there are also local rules that might vary in terms of what kind of media and under what conditions media are available to the incarcerated individuals that vary depending on security levels. During the history of the modern prison (from the mid-nineteenth century onward), shifting penal ideologies have sought to manage communication and media use in different ways. The early prison regimes tried to minimize the incarcerated individuals' communication with others to create "self-communication," or introspection, for the incarcerated to ponder their crimes and develop into reformed individuals. Media use was restricted to an absolute minimum and normally only consisted of the Bible and a few other books that were seen as morally constructive. As rehabilitation and normalization grew stronger and penal ideology (as well as practice) changed in the twentieth century, communication was less restricted: on the contrary, the road to reform went through making the incarcerated communicate (with each other, with staff, and with the surrounding world). Media use went from being forbidden to becoming encouraged as part of the treatment and care of the incarcerated (access to books, newspapers, radio, music systems, and later television became a right in prison). "Communication" became a positively charged buzzword within the developing penal regime from the mid-twentieth century and still is today. Even so, media access within the prison is a sensitive issue that might spark popular debate.

Yvonne Jewkes and Bianca Reisdorf argue that "media technologies fundamentally challenge the historical meaning and functions of the prison as an archetypal 'closed,' sequestered, and restricted space that assaults self and personhood" by opening up the space of the "total institution" to the external world in terms of both providing alternative realities and allowing for connections with people beyond the prison walls (Jewkes & Reisdorf, 2016, p. 536). At the same time, smart digital technologies including touchscreen kiosks on which incarcerated individuals can arrange family and legal visits, order certain products, and choose their canteen food contribute to a shift in governance of prison populations. Jewkes and Reisdorf (2016) argue that these technologies are part of a refashioning of power and control, moving from a coercive model to a model that is focused on self-interest and self-regulation, or what has been described as responsibilization. Initiatives such as the PrisonCloud implemented in several Belgian prisons that provide in-cell digital services, including telephone calls and service ordering via a centralized personal digital platform, have been shown to contribute further to isolation of incarcerated individual within the prison while allowing for mediated connections beyond the prison walls (Robberechts & Beyens, 2020). However, there remains a strong digital divide between prisons and the outside world as well as between prisons that handle access to digital media technologies quite differently. In most jurisdictions access to digital media is still strongly restricted, and investments are made to block external communication from the prison space (Van de Steene & Knight, 2017).

Public discourse often emphasizes the importance of a "total institution" that controls and secludes the criminal subject from society as the preferred mode of penalty, while access to (new) media technologies is widely regarded as a privilege (Jewkes & Johnston, 2009). In that sense, access to media of all kinds but especially of digital media technologies in the prison space potentially runs counter to common ideas of incarceration that are based on isolation, solitude, retribution, hardship, and suffering. For these reasons, the question of incarcerated individuals' access to (digital) media has become an ideological battleground. At the same time, advocates for incarcerated peoples' rights argue that the absence of computers and digital technologies in the everyday lives of incarcerated individuals impedes rehabilitation and comes close to a form of censorship. This reduces them to second-class citizens and should be considered as similar to other forms of deprivation such as low income, unemployment, and inadequate access to education (Jewkes & Reisdorf, 2016). Peter Scharff Smith (2012b), for instance, argues that internet access should be considered a human right; this claim has become part of the normalization of prison life that informs the dominant approach in Norway, Sweden, and Finland as well as Denmark and Iceland. The normalization approach emphasizes that prison life should resemble as much as possible life outside of the prison walls, including stable structures of everyday life such as work and leisure activities. The approach is furthermore expressed in collegial relations between incarcerated individuals and guards and in a commitment to providing reliable social services, including education and job training, to incarcerated individuals and staff (Reiter et al., 2018). To some extent, the normalization thesis coexists with the narratives of surveillance and control: new smart products and solutions are often marketed as giving new opportunities for rehabilitation (e.g., through extended possibilities for incarcerated individuals to communicate with friends and loved ones) and at the same time deepening the possibilities for control and surveillance (e.g., through automated monitoring of calls and messaging systems that allow the incarcerated to keep in touch with people outside the prison through text messaging).

APPROACHING PRISON MEDIA: A NOTE ON METHODS

The German company Gerdes offers a platform solution called PrisonMedia. This platform has been implemented at Finland's first smart prison, which opened in February 2021. However, prison media is nothing that you find in an online store for purchase.[3] Instead, prison media is a notion that we actively produce through our methods and theoretizations and has a social life in itself (Law & Ruppert, 2013). Prison media is a heuristic that aims to make visible how the development of the prison is entangled with the development of our media; it is a way of seeing, and this way of seeing is facilitated in different sights where we trace the overlaps of the prison and media infrastructure. These sights include archival sources documenting the Swedish Prison and Probation Service (including order books, client registers, documents by the architectural committee, official statistics published by the service since 1948, reports by the service's department for work operations from 1958 to 1962, and reports and concept papers by the board of the service including annual reports from 1970 to 2000). Additionally, we have submitted two Freedom of Information requests to gain an overview of current external clients as well as orders of goods and services by the Prison and Probation Service. Besides analyzing these documents, we have conducted field visits at three out of seven high-security prisons (Security Class 1) in Sweden, including visits of the prison workshops. The visited facilities house between 171 and 405 incarcerated individuals. We also track prison media through in-depth interviews with the work operation coordinator at the headquarters of the Prison and Probation Service and managers of work operations at the visited high-security facilities as well as informal conversations with workshop leaders, amounting to in total four interviews during which we took extensive notes. In addition, we attended two security technology fairs that exclusively or partly cater to the corrections sector, one in the United Kingdom and the other in the United States. At the expos, we collected materials on technology companies that are active in the corrections sector by interviewing the sales managers as well as collecting information materials provided at the exhibition stalls. Furthermore, we attended the European conference "Technology in Corrections" held in Lisbon in April 2019. The intersectorial meeting is jointly organized by the European Organisation of Prison and Correctional Services and the International Corrections and Prisons Association and brings together corrections-sector practitioners, industry representatives, and academics. The participant observation during the conference was documented in extensive field notes, including PowerPoint presentations that were made available for free after the conference. Additionally, we have received access to transcripts of semi-structured interviews with 71 incarcerated individuals and 126 staff members at Swedish prisons including wardens, guards, and educators as well as work operation managers. To be able to reanalyze the material, we received approval by the Swedish Ethical Review Authority in 2019. The interview material was produced by the Swedish National Council for Crime Prevention. For chapter 5, the last empirical chapter on prison tech, we also delve into patents and patent citations as source material. In these diverse materials, we trace and make visible prison media as both devices and discourses.

The prison also invokes very concretely the role of boundaries between the normal and the deviant. Prisons seemingly establish clear borders between the inside and the outside through physical markers of the perimeter fence, the high walls, the security locks, and doors as well as through the strict regulation of communication. But is the boundary between the inside and the outside that clear cut? Prison media as heuristic and material artifact can be considered as boundary objects, a notion established by Susan Leigh Star and James Griesemer (1989). Their starting point is that there are different social worlds following their own rules, codes, habits, and ways of making sense. These separate social worlds are, however, connected through boundary objects that travel between them. Boundary objects are enacted and made sense of in different ways across social worlds, but they are fuzzy enough to be recognized in different contexts. They need to be unspecific to a certain degree to flatten potential tensions between the social worlds and their different ways of sense making. Through their mutual recognition in different social worlds, boundary objects allow for collaboration across boundaries and thereby make them less absolute. Prison media—media produced for and in the prison as well as the prison as medium—can work as such boundary objects. They travel between the prison worlds and other social worlds and blur the boundaries.

Star and Griesemer developed the notion of boundary objects originally in the museum context, where theorists and amateurs collaborate, and highlighted several boundary objects that facilitate this collaboration (repositories, ideal types, coincident boundaries, standardized forms). Prison media are similarly a collection of boundary objects: they include practices of work, architecture, and technology development. While we will highlight each of these features in the chapters to follow, we also need to engage with the movement of prison media between different juridical systems. Although much of the material we rely on emerges from the Swedish penal system, prison media have international layers including transnational, globally active companies, and other entities that dominate, for example, the prison tech sector. There are also entanglements across different countries and penal systems that are constituted by specific national jurisdictions and regulations but also depend on local regulations that vary from facility to facility emerge in and through prison media. While the systematic study of archival material has been focused on the Swedish context, we also draw on examples from other countries, illustrating the international prison media complex (we mainly discuss a European and North American context). The international cases situated outside of Sweden are telling examples representing cutting-edge projects that strongly illustrate our points or describe technologies and related developments that have implications on a global scale. This eclectic approach leads to shifts between different penal systems that range from Scandinavian exceptionalism to mass incarceration in the US context, which also illustrates the changing character of the prison media complex where entanglements between state institutions and private corporations emerge on local, national, and global levels. At the same time, prison media emerge in the context of supranational reference frames such as international agreements on human rights, including those for incarcerated individuals and monitoring by supranational bodies such as the United Nation's Committee against Torture and the Council of Europe's Committee for the Prevention of Torture and Inhuman or Degrading Treatment or Punishment (also known as the Committee for the Prevention of Torture) that have repeatedly criticized the extensive use of solitary confinement, including complete exclusion from any kind of news media at Swedish remand prisons.[4]

In the chapters that follow, we mobilize the notion of prison media to explore the associations and ways of ordering they allow for. We do so by exploring them as infrastructures that are relational and constituted of layers and illustrate the complexity of how prison worlds and social worlds are related, but first we establish a brief time line of the modern prison regime with a specific focus on the Swedish context in chapter 2.

OUTLINE OF THE BOOK

This book is divided into six chapters, which together develop the notion of prison media both empirically and theoretically. Chapter 1 has introduced the conceptual entry points and motivation for this investigation. Here we answer the question of what prison media are and why we should care about them.

Chapter 2 presents a short historical overview of shifting penal regimes in Sweden while connecting to global developments concerning ideas of punishment and reform. Consequently, the chapter develops a time line of penal regimes. We discuss central ideas of punishment and rehabilitation and how they have changed over time depending on shifting political context. While the historical investigation of chapter 2 largely broadens the scope of the book, we still center on the role of media and communication infrastructures as being fundamental to the evolution of different penal regimes.

Chapter 3 engages with the history of media work among incarcerated people from the twentieth century to the present. Work and productive labor are central aspects of prison life. Chapter 3 takes on prison media work and explores how incarcerated individuals were engaged in the construction, maintenance, and repair of infrastructures of communication and media throughout the twentieth century. The analysis of media work by incarcerated individuals problematizes and enriches ongoing theoretical debates on media work and labor within media studies as, for example, developed by Hesmondhalgh and Baker (2011). Throughout the chapter we relate prison media work to broader structural questions, such as precarity and exploitation in contemporary capitalism as well as media work that has been investigated earlier as hidden work of the media industries (Mayer, 2011). Another important aspect of prison media work is the role of prisons as test beds for new technologies of surveillance and control. The "work of being watched" (Andrejevic, 2002) that incarcerated persons perform is a key dimension of prison media work but also increasingly relates to societal transformations in labor. Here, we also relate the work of incarcerated individuals to the media work of prison guards. Not only is much of guard work about controlling, moderating, suppressing, and steering communication and media practices (who talks to whom, what media content can be consumed by whom, etc.), but much of the work that guards do also relies on and has been crucial in forming now-ubiquitous media technologies, such as CCTV systems and communication radio.

Chapter 4 focuses on prison media architecture. Prison architecture and construction since the birth of the modern prison in the nineteenth century have been deeply concerned with questions of communication: how to allow for or deny communication between different parts of the prison population or between incarcerated persons and the outside world. Incarcerated people themselves have used prison infrastructure tactically to make it work for their communicative purposes. Chapter 4 considers prison architecture itself as a medium regulating, allowing, and constraining media distribution that includes, along with other media such as the television and the computer, walls, fences, gates, hallways, wings, and cells as well as the sounds and smells that structure time, space, and communication for incarcerated individuals, guards, and visitors (Fassin, 2017). In addition, chapter 4 analyses how architects and officials historically imagined the prison complex and also engages with the question of how the prison functions as a means for controlling movement and communication, highlighting how the prison as infrastructure has been used by incarcerated individuals in sometimes unexpected, daring, and defiant ways for unsanctioned communication. Empirically, chapter 4 concentrates on prestigious and central prison complexes with the highest security level (Security Level 1) in Sweden, which stand out in terms of their architecture (e.g., the prisons Hall, Kumla, and Tidaholm), and examines how their design draws on international standards and discussions of prison security as well as manifests central ideas of penology.

Chapter 5, the final empirical chapter, zooms in on technology development in the prison context. We argue that prisons serve as test beds for technologies that later spread, with slight modifications, beyond prison walls and hence have direct repercussions for society in general. Focusing on the example of ankle monitor systems, the chapter traces the history of specific prison media technologies through patent citations. First developed in the 1960s and then tested in the 1980s in the US, ankle monitors are now used widely in numerous penal regimes to control offenders under house arrest or on parole. Ankle monitors have consequently been envisioned as an alternative to cost-intensive, physical prisons, contributing to what could be called "mobile incarceration" outside of prison buildings. In chapter 5, we argue that monitoring devices for tracking defendants have been remediated and popularized in the form of not only surveillance devices (e.g., for children) but also activity tracking devices for self-monitoring (such as wrist-worn trackers and smartwatches). Chapter 5 demonstrates the crucial and direct relationship between prison technologies and technologies developed for communication and control. We argue that prisons, both historically and currently, function as test beds for technologies that later make their entry into commercial and popular use with only slight modification. We highlight the role of prisons for technological development and dominant sociotechnical imaginaries that crystalize in the prison context.

Chapter 6 integrates the book's three empirical sites—work, architecture, and technology development—to further theorize the notion of prison media and discusses their implications for media infrastructures more generally. By way of concluding, we argue that we are increasingly living in a culture of mobile incarceration emerging both in the punitive strategies that move out of the prison walls, dispersing punishment across multiple sites, and also and more importantly in data-driven algorithmic culture that fosters self-optimization and surveillance. Accordingly, the individual is increasingly incarcerated by technologies of the self and disciplinary power of networked, digital media.

2

PENAL REGIMES AND PRISON REFORMS: 1800–2000

Prisons as we know them are a product of modernity. Born in the late eighteenth and early nineteenth centuries, the "modern prison," or the "prison system," is intertwined with the social organization of capitalism, liberal ideas and ideologies of the individual and of social change, and modernist architecture and social engineering. We take our starting point in the nineteenth-century prisons, but the emphasis in the empirical chapters is on the 75 years between the end of World War II in 1945 until the present. This is the era of the welfare state, advanced capitalism, the industrialized prison, and modern electronic and digital (mass) media. The history of the modern prison can be told in many ways, and the literature covering the various aspects of this history is vast. We can here only give some brief accounts and highlight a few important points in the history of the prison system that are of relevance for our understanding of prison media. The developments of the Swedish prison system follow, in some respects, broad patterns in the global North. However, there are also national and cultural specificities.

The Swedish prison system has often been understood as a specific form of penal regime influenced by the Social Democratic welfare state. Historically, it has been described as more humanitarian and with a strong focus on normalization (i.e., the belief that life in prison should resemble ordinary life outside of the prison as much as possible); additionally, Nordic societies have often been perceived as having a comparatively small prison population (Lappi-Seppälä, 2007; Pratt & Eriksson, 2012). The penal "exceptionalism" of the Nordic countries has been a topic in both scholarly and public debate for a long time, but during the last decade this image is slightly changing. While the influence of global neoliberal penal ideologies may be to blame for the waning of Nordic exceptionalism, it is also possible that said exceptionalism was never fully achieved (Scharff Smith, 2012a). Although a full account of these debates is beyond the scope of this chapter, we flesh out some of the specificities of the prison in relation to the Nordic welfare state.

At the same time, we situate the history of the prison system in relation to broader and more general developments, particularly in Sweden but with an international outlook.

THE BIRTH OF THE MODERN PRISON

Although different penal regimes have succeeded each other throughout history, the modern prison is a product of the late eighteenth and early nineteenth centuries. As capitalism emerged alongside the Enlightenment movement in the eighteenth and early nineteenth centuries, corporeal punishment and work as a form of punishment were called into question. Instead, criminals should be punished for their actions in a way that allowed for reforms through introspection, soul-searching, and religiously inspired repent. This stems from a new understanding of the individual as a possible object for "reform." The individual in general was seen as "formable," and hence through proper treatment even the worst criminals could be bettered. Furthermore, "doing time" was in a new capitalist economy increasingly understood as a form of punishment, as time had been reimagined as a currency: time is what a wage laborer exchanges for a wage. Time itself has a value, or a prize, and to confiscate someone's time through incarceration can thus become a punishment (Melossi & Pavarini, 1981). These general lines of thought crystallized in the modern prison.

Prior to this period, Scandinavian prisons were both brutal and ill-developed. They were harshly criticized in the writings of international penal reformers such as John Howard (Smith & Ugelvik, 2017). In Negley K. Teeters's work in comparative penology titled *World Penal Systems* (1944), the Scandinavian prisons of the eighteenth and early nineteenth centuries are cited as an example of "the most barbaric forms" of punishments. He states that the Scandinavian countries were the most "bloodthirsty and severe" in Europe at the time (Teeters, 1944, p. 86). At the turn of the nineteenth century, however, there was a general and international shift in how Western society looked upon crime and punishment, a shift from corporeal punishment to surveillance and self-regulation that, as Michel Foucault famously argued in *Surveiller et punir* (1975), transformed the penal system and ideology. Corporeal punishment was gradually abandoned. Instead, incarcerated individuals were held and surveyed in institutions meant to enhance their moral reform and capacity for self-reflection and surveillance. Activists and reformers advocated for these changes, regarding them as a necessary humanization of the penal system. Their ideas found support among social elites rather than the general populace (Smith & Ugelvik, 2017).

Enacting these changes required an international shift in the organization of punishment in most Western countries. Modern Swedish prisons were inspired by new models imported from the US such as the Philadelphia model, also called the "solitary system," and the Auburn prison, known as the "silent system." The Quaker vision of penology—the Philadelphia model—was first introduced and reflected in the architecture of the Eastern State Penitentiary in Philadelphia. The prison architecture was developed based on the premise of solitary confinement that should allow incarcerated individuals to reflect about their wrongdoings in total isolation. Individual cells were lined up in a wagon wheel construction in which cell wings radiated in a circle from a tower in the center to allow for constant surveillance (figure 2.1). This architectural model of the Eastern State Penitentiary was exported globally, and more than 300 prisons were constructed in similar ways (Johnston et al., 1994).

In contrast, the Auburn system was based on the maxim of silence but gradually loosened up isolation. While solitude was only kept at nighttime, incarcerated individuals were allowed to do some work during the day. They were working either alone in their cells or in communal workshops at daytime on the condition that they kept absolute silence. The Auburn system gradually replaced the Philadelphia model in many places, as it promised to contribute to the rehabilitation of incarcerated individuals by establishing personal discipline and respect for the value of work and other people. One of the typical prisons following the Auburn system is Sing Sing Prison in New York state. The architecture of Sing Sing reflects the full integration of work into the daily lives of the incarcerated people and includes workshops and prison factories. The Auburn system was considered revolutionary because it was supposed to return profit to the state, and work was fully integrated into the routines of the prison.

The control of movements, sounds, and communication was an important feature of the modern prison and made strong impressions on seasoned officers of the correctional services as well as on visitors, as two encounters illustrate. The famous Danish author Hans Christian Andersen gave the following account of the mixed emotions of architectural grandness and enforced silence in one of Sweden's cell prisons in the 1840s:

> Like a great castle, this building—whitewashed, smiling, with windows on windows—is located in beautiful nature, next to a small river just outside the

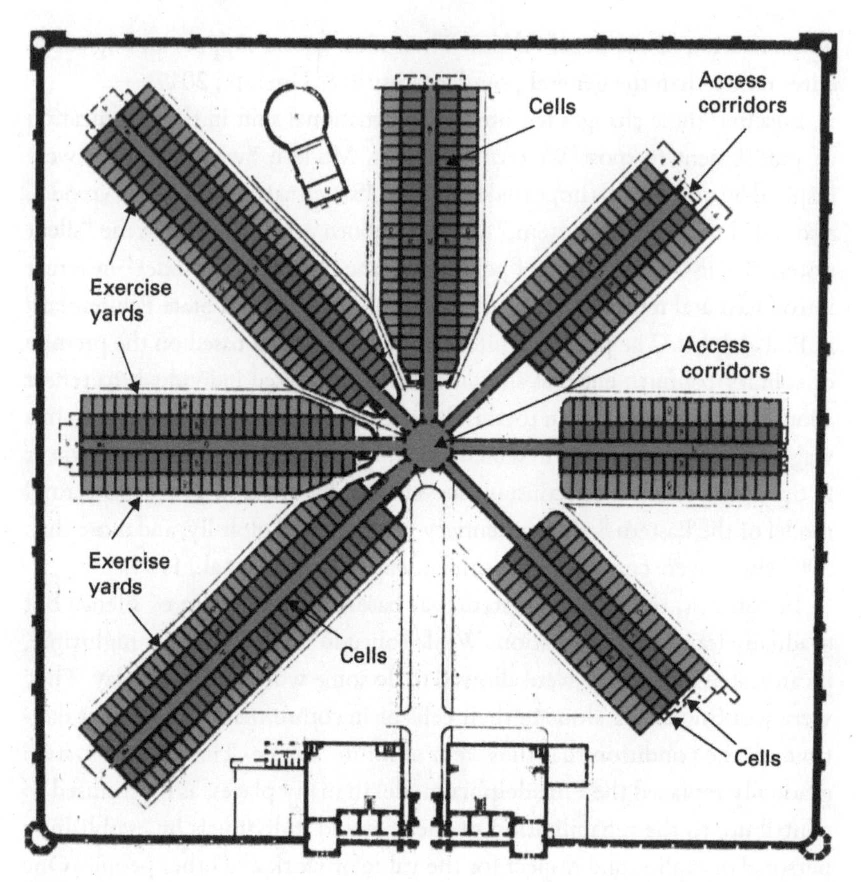

Figure 2.1
The wagon wheel architecture of the 1836 floor plan of the Eastern State Penitentiary in Philadelphia, Pennsylvania, showing access corridors (the "spokes" of the wheel in the diagram), cells (inner areas in dark gray), and exercise yards (outer areas in light gray). *Source*: Myles Zhang, Attribution-ShareAlike 4.0 (CC BY-SA 4.0)

city center. But soon one discovers that there is a grave-like silence in this place. It is as if no one lived here, or as if the building were abandoned in the time of the plague. . . . The whole thing is a well-built machine, a nightmare for the soul. In the door of each cell is a piece of window-glass as large as an eye. A flap outside covers it, and here the prisoner-guard, unnoticed by the prisoner, can see all that he does. But the prison-guard must come quietly, silently, because the prisoner's hearing is in the loneliness strangely sharpened; I slowly turned the flap and my eye looked into the obscured room where the prisoner's eyes instantly met mine. It is airy and clean in there, but the window sits so high that it is impossible to look out from it. A high stool that is stuck as a kind of table and next to it a bed that can

> be hung up with hooks in the roof is the whole furnishing. . . . Outside, in the healthy sunshine, there is full activity, but in here it is always as silent as midnight. The spider that spins down the wall, the martlet that flies close enough to be seen from the window, even the stranger's steps in the corridor outside the cell doors is an event in this monotonous, mute life where the prisoner's thoughts roll into themselves. (Andersen, quoted in Lundberg, 1997, p. 12)

Similarly, Torsten Eriksson, who became general director of the Prison and Probation Service in 1960, describes his first day as an assistant to the warden of Karlskrona prison in 1930, when the prison buildings from the end of the nineteenth century were still in use as follows:

> Every now and then the guards would sneak to the cell doors and look inside through the little peephole to check on the prisoners. The system was built so that the prisoner could always and everywhere be observed. Orwell's future vision of 1984 with Big Brother's television eye watching in every room and on every street corner was already a reality in the prison world. (Eriksson, 1967, p. 9)

Even if modern cell prisons were perceived as a humanization of the penal regime, the problems and cruelty of a system based on "solitude" and "silence" quickly became obvious. The mental health of the incarcerated individuals was severely damaged. In response, prisons in Western countries began to take steps to limit isolation and introduced more communal elements. Scandinavian prisons, however, continued to use solitary confinement for a much longer time than in the rest of the Western world. Even though changes and reforms of the system began in 1906, it was not until 1945 that the so-called *ensamhetsstraffet* (punishment in solitude) was abandoned in Sweden. Today, solitary confinement remains the norm for remand prisons for which Sweden has been continuously criticized by the United Nations Committee against Torture as well as the European Committee for the Prevention of Torture and Inhuman or Degrading Treatment or Punishment, part of the Council of Europe (Smith & Ugelvik, 2017).

POSTWAR PENOLOGY AND THE INDUSTRIALIZATION OF THE PRISON

In 1940 the opening of Hall, a new modern prison outside the town of Södertälje that is also the main arena of the novel *Grundbulten* (The foundation bolt), marked a new era in Swedish penal history, namely the emergence of the postwar penal regime. The architecture of Hall was supposed to facilitate a communal life and at the same time represent an "industrialization" of prison life in which work became more central to the everyday existence within the prison walls. It was hence a material manifestation of a new penal ideology that took shape in the postwar period. The juridical manifestation of the same shift in penal ideology was the Implementation of Sentence Act passed in 1944. This act laid great responsibility on prisons to reform the incarcerated individuals to a "socially useful life" (Prison Instruction 1938, quoted in Nilsson, 2013, p. 41) after their prison sentence was served. Some of the paragraphs from the 1945 law can serve as an example of the shift in penal ideology, including the rehabilitative character of work:

> § 24 Punishment shall be enforced in such a way that the inmate's correction is promoted. He shall be employed with appropriate work.
>
> § 25 The inmate should be treated with respect for his human dignity. Harmful effects of the loss of freedom must be prevented as far as possible.
>
> § 29 Gymnastics and sports exercises shall be arranged for the inmates when possible.
>
> § 30 The inmate should be encouraged to [engage in] studies and other suitable leisure activities.
>
> § 31 To the extent that it can be done without inconvenience, the inmate must be able to acquire or receive books, magazines, newspapers and other things that can facilitate for suitable activities during leisure time. Opportunities for newspaper reading must be prepared.
>
> (Implementation of Sentence Act, 1945)

This law formalized changes that had been under way since the beginning of the early twentieth century. The director generals of the Swedish penitentiary authorities had worked to reform the prison system by implementing policies promoting more humane treatment for incarcerated persons and more possibilities for decent work and better living conditions within the prisons. Following the democratization in the 1920s, Social Democratic and Liberal governments were elected. They had a different approach to prisons and penal policy than the governments from the predemocratic era. In the 1930s the Social Democratic attorney general, Karl Schlyter, presented a range of ideas for reforming the prison system under the slogan of "Depopulate the prisons." These political struggles came together in the postwar prison reforms, which also contributed to the larger project of forming a welfare state (Smith & Uglevik, 2017). The new penal ideology was characterized by many of the core principles common to postwar welfare states, especially the notion that it is the state's responsibility to care for its citizens, including its criminals, by providing rational, pedagogically, and socially responsible institutions. More specifically, the new prison system's emphasis on preparing incarcerated individuals to reintegrate into society as sound and productive individuals conferred a new status on prison work that, as prison historian Roddy Nilsson has shown (2013), supported the strong moral charge given to work in Social Democratic welfare states. The social "role model" par excellence during this period was, according to Nilsson, the "loyal and conscientious working-class man who contributes to the welfare of the whole society" (Nilsson, 2013, p. 43). As a result, work became the principal treatment method in the welfare state prisons.

This shift in Swedish penal ideology corresponded with broader international trends and more general shifts tied to a changing Western political and economic structure. In Sweden, the postwar period witnessed rapid economic development; likewise, the Social Democratic government, with its Keynesian economic policy, made fighting unemployment its main objective. At the same time, the rise of late or so-called monopoly capitalism and a new importance of large-scale bureaucracies (both within the state and within commercial enterprises) further incentivized the industrialization of the prison. The penal system's new focus on work and especially industrial work served these emerging state and commercial bureaucracies with cheap labor, delivering the goods needed to create and maintain the bureaucratic corporations and the welfare state of postwar Sweden. Furniture and equipment for schools, universities, libraries, hospitals, public authorities, and the state apparatus as well as for new communication technologies such as the telephone system, broadcasting, and traditional communication infrastructures such as the postal system were, as we will see, delivered by prison labor.

In combination with the mandatory industrial work often assigned to incarcerated individuals, postwar prison architecture sparked a critical debate on penology in the mid-1960s. This discussion was part of a broader contemporary debate over how the welfare state treated its outcasts. The mentally ill, drug addicts, sexual minorities, and other groups organized in associations and unions to highlight the inhumanity of certain welfare state institutions. The debate was international and tied to general political changes in 1960s, when a progressive, left-wing, and liberal critique of the welfare state was formulated by a younger generation and among academics.

Most famous regarding prison reform is perhaps the French Le Groupe d'information sur les prisons that was active in the early 1970s and that both Gilles Deleuze and Michel Foucault participated in. Foucault's classic study of punishment was in many ways a direct result of his experiences as an advocate for prison reform in France.

In Sweden, a similar organization called Riksförbundet för Kriminalvårdens humanisering (National Association for the Humanization of Corrections, KRUM) was formed. For a decade, KRUM played a key role in the political discussions of the Swedish penal system. KRUM was an association made up of liberal and left-wing academics, students, and social workers together with the incarcerated themselves. They organized strikes and took part in the formation of labor unions within prisons, published numerous books and articles on the state of the Swedish prison system, and lobbied for reform of the penal system. Popular novels and films were released questioning the very idea of the prison, and from the Left to the Right there was a relative consensus on the need for liberalization and reform of the Swedish penal system. For example, the attorney general in the 1970s, a Social Democrat, stated that Sweden should have a maximum of 500 incarcerated individuals (the actual number was at the time around 4,000). Similarly, the leader of the conservative party was an activist for a more humanitarian penal policy and was personally involved as a volunteer probation officer for recently released criminals.

Socially and economically, these pushes for penal reform coincided with historically low numbers of unemployment. This is unsurprising, as the number of unemployed in a society has been shown to correspond with the size of the prison population. According to Georg Rusche and Otto Kirchheimer (1939/2009), for instance, the supply of labor determines the design of the prison system. As such, punishments tend to become harsher when there is a surplus of labor and vice versa. This materialist theory of punishment has found empirical support, even though later developments of the theory tend to underline the need to combine the materialist explanation of punishment with social, political, and ideological factors (De Giorgi, 2006).

Changes in the Swedish penal regime at this time coincided with other forms of economic change such as globalization and the beginnings of outsourcing of labor, which made cheap prison labor less attractive. Finding work for the incarcerated became increasingly difficult during the 1970s. The future of the prison in its current form looked gloomy; indeed, for many the prison as an institution came to look like a parenthesis in the history of punishment, soon to be abandoned altogether. Prison theorists Dario Melossi and Massimo Pavarini (1981/2018), for example, predicted that "the post-industrial era will accommodate ways other than imprisonment for social control and discipline," suggesting that "propaganda, mass media, [and] a new and more efficient network of police and social assistance" would replace incarceration in the future (p. xxxi).

In 1974, the new Prison Treatment Act (Swedish Code of Statutes, 1974, p. 203) was passed following a long period of parliamentary discussions and reports on the subject. This new law brought the modern organization of the Swedish Prison and Probation Service into being. Public authority was reorganized, and several of the "democratizations" pushed for by KRUM and others were realized. For example, incarcerated persons' rights to organize themselves and to be consulted by the prison directorate on issues on policy and administration were expanded. There was also a new focus on "noninstitutional care" that allowed for people to serve their time outside of prison. The law also represented an effort to pay incarcerated individuals market salaries for their work and to better facilitate their access to education, which meant that new professions, such as youth workers and educators, entered the prison institutions (Tham, 2001). Melossi and Pavarini (1981) have argued that at moments when organized labor wields more economic power than capital, prison sentences are pushed back, and the prison conditions are improved. The 1970s in this respect probably represents the peak of the power of the labor movement in strength and organization throughout the West and especially so in Sweden, where the Social Democratic Party received over 50 percent of the votes in the election in 1969, the highest percentage it had ever received. Toward the end of the 1970s the penal regime changed once again, initiating a new era of mass incarceration.

NEOLIBERAL MASS INCARCERATION AND PRISON DECENTRALIZATION

If the prognosis in the 1970s had been that prisons were soon to be consigned to the dustbin of history, the reality of prison development has completely upended that assumption. Instead of shrinking, prison populations have grown rapidly in many parts of the world. The length of sentences has increased, and punishment and incarceration have been given more weight than earlier values such as reform and rehabilitation.

The period from the 1980s until today has been much discussed as an era of mass incarceration. The prison population has grown almost exponentially, especially in the United States. The US is by far the world leader in incarceration, with 655 imprisoned citizens per 100,000. Sweden is often put forward as the opposite of the penal regime in the US. Together with the other Nordic countries, Sweden is among the countries with the lowest rate of imprisonment in the world (World Prison Brief, 2017, https://www.prisonstudies.org). As discussed by Michael Cavadino and James Dignan (2006), the Swedish penal system has been famous for relatively good prison conditions, decent staff-inmate ratios, few instances of riots and disorder, few escapes and escape attempts, and a generally high level of legitimacy in the eyes of the public (Cavadino & Dignan, 2006, p. 159). Since the late 1980s and early 1990s, however, the reality of this assumption has changed to some extent. For example, the rate of imprisonment in Sweden grew steadily from the 1980s until the mid-1990s. The public debate on crime and punishment has also become marked by calls for greater use of imprisonment, and sentencing scales have been tightened for many offenses (von Hofer 2003). In the words of Cavadino and Dignan (2006), "the 'penal Zeitgeist' in Sweden is moving in the same direction as in most of the West" (p. 159). There are different explanations for this fact.

First, this shift corresponds to changes in Sweden's political and economic structure. The end of a Keynesian paradigm and the triumph of neoliberal economics has precipitated a global spike in unemployment rates. Neoliberal economic theory rests on the twin premises that combating inflation is a central virtue and that to keep inflation down, a certain amount of unemployment is necessary: because unemployment is understood to restructure the power balance between capital and labor in favor of the former, a certain rate of unemployment helps to keep wages down. The prison functions as a social institution that absorbs the "negative externalities," that is, the social costs of the economic system; thus, as Rusche and Kirchheimer (1939/2009) showed as early as the 1930s, unemployment rates correlate directly to the size of incarcerated populations.

Second, an ideological shift has accompanied the above economic shift. In the late 1970s, populist and right-wing politicians began to push agendas claiming to be "tough on crime," tilting the penal debate in favor of longer sentences, the criminalization of a broader range of behavior, and incarceration as a punishment for more minor crimes than ever before. Globally, this rhetoric has been used by politicians from Ronald Reagan and Margaret Thatcher in the 1980s to the centrist Social Democrats of the "third way" in the 1990s to advance political programs meant to undermine the welfare state, such as lower taxes, cuts to social benefits programs, and tougher penal policy. In the Swedish context, "rehabilitation" has generally been seen as the main purpose of prisons throughout the twentieth century. Social democratic ideology, in line with the "positivist" school within criminology, forged a consensus among the penal elite about a strongly rehabilitationist penal code (Cavadino & Dignan, 2006, p. 155). However, this consensus started to break down in the late 1970s. An important mark in this development was the report *A New Penal System: Ideas and Proposals* (1977), authored by a working group set up by the National Swedish Council for Crime Prevention. This report, written during a period of right-wing government in Sweden, introduced the "justice model approach," which dictates that sentences and sanctions should be determined not by the possibility of rehabilitation but instead by the penal value of the offense. In 1988, the ideas put forward by this working group resulted in a new law, the Sentencing Reform Act, which stated that sentence length should be determined "with special regard to the harm, offence or risk which the conduct involved." Especially among right-wing politicians and parties, this was seen as an opportunity to demand harsher sanctions (Cavadino & Dignan, 2006, p. 157). During the following years, an economic crisis in the early 1990s and a right-wing government brought penal issues to the forefront of policy development. In response to a policy document called *To Restore a Degenerated Criminal Policy* published by the conservative minister of justice Gun Hellsvik in 1993, sentence lengths were increased for over 20 different crimes, and Sweden's prison population grew significantly (Jareborg, 1995).

Privatization, a third feature of the neoliberal landscape since the 1980s, may also be responsible for the sharp rise in incarceration over the last four decades. Toward the end of the 1970s, the Swedish society pushed for the decentralization of large bureaucracies, public authorities, and public services. In some ways, this decentralization was a response to the progressive critique of the welfare state and its calls to democratize and humanize welfare institutions, allowing citizens more freedom of choices instead of social engineering implemented by the state. Prisons were among the institutions that became increasingly decentralized in the 1980s. In Sweden, this decade marked a new era of prison constructions. During the 1980s, 15 so-called local prisons were built across the country. This was an attempt to decentralize the Swedish Prison and Probation Service and make it easier for incarcerated individuals to serve their sentence close to their homes and in regular interaction with the local society outside the prison walls including friends and family, local enterprise, and employment and social services. Later developments with ankle monitors pushed this tendency of decentralization even further when shorter sentences could be served from incarcerated individuals' own homes.

In many other ways, however, the push for decentralization was also meant to open the large welfare systems for private investment and private capital, a victory for free market ideology and a way for capitalist interests to take control over the previously de-commodified sectors of public welfare. In Sweden and internationally, hospitals, elderly care, schools, childcare, and other institutions were privatized in the 1980s. However, in Europe if not elsewhere, prisons, like many other aspects of the oppressive state apparatus including the police and the military, have not been privatized to the same extent as other social services. European governments retain some skepticism regarding trusting the enforcement of state violence through incarceration to private interests. Countries such as the US, the United Kingdom, Canada, Israel, South Africa, and New Zealand all have a private prison industry. The privatization of prisons must be seen as an important context for expanding prison populations, at least in some parts of the world, as the growth of numbers of incarcerated individuals and the increased number of criminalized offenses becomes a political issue as well as an economic interest for the developing prison industry.

Not only has the period from the 1980s until today seen a growth in prison populations, but the shift in penal ideology in society has also meant that the content of punishment has been gradually transformed. The "progressive" reforms of the 1970s intended to help rehabilitate incarcerated individuals have been gradually replaced by more pessimistic assumptions about the prison's capacity to effect positive change. The contemporary penal system imagines incarceration to protect citizens from incorrigible criminals. This shift began in the 1980s but was accelerated during the 1990s, when the public debate on Swedish prisons mainly focused on the need to increase sentences and to build new and more secure superprisons to house people. Not only does the contemporary prison downplay the value of educating and reforming incarcerated individuals, but work has also come to play a smaller role in the prison ecosystem. Global trends toward automation and the increasing outsourcing of labor have made the demand for cheap prison labor drop. It has been increasingly difficult to find customers for the industrial production conducted in prisons (Kindgren & Littman, 2015).

COMMUNICATION AND MEDIA IN SHIFTING PENAL REGIMES

As the prison itself is a structure that organizes communication (i.e., movement in space, interactions, and modes of communication) and for these purposes makes use of different media technologies, the different penal regimes outlined above also—to some extent—entail different ideas and ideologies of communication. The first modern prisons of the nineteenth century were often associated with the principle of the panopticon; that is, all surveillance takes place from one central node in the prison facility in a process in which information (about the behavior of the incarcerated) is transmitted to the guards. But in this centralized communication system, information could also travel the other way. While the modern prisons of the nineteenth century had a regime of strictly prohibited interaction with other fellow humans, they privileged communication as a form of "broadcasting": incarcerated individuals could receive messages broadcast from one central point, usually a lectern installed between the balconies where the cells were located, from which the priest or the warden could address all incarcerated individuals at once. The centralized and strictly controlled media world of the modern cell prison also included some rare media texts chosen by the central authorities such as the New Testament and a handful of books. The early prisons were by no means places of noncommunication even though they were media environments characterized by scarcity. On the contrary, the early modern penal regime put strong emphasis on the transformative power of (mediated) communication: when the right messages were received at the right time, they would help reform incarcerated individuals and transform them into a socially benign existence.

In the industrial postwar penal regime, a new communicative ideal took shape. As discussed by John Durham Peters (1999), the postwar period in general tended to rearticulate a host of issues as problems of communication.

If the media and communicative environment of the modern prison had been one of scarcity and "broadcasting," the 1950s and 1960s was an era in which prisons were organized according to the "small group principle" in which interaction between incarcerated individuals and their communication with each other was understood as a key element in their reform. The policy documents from the era discuss the importance of "trust" and the use of "motivational conversation" between staff and incarcerated. Furthermore, a range of different communications media and cultural forms now entered the prisons: radio, film, newspapers, theater, music (both recorded and live), and later television. The strong emphasis on "normalization" within the Scandinavian prison system in this period meant that media use—reading books, newspapers, and magazines as well as listening to the radio and going to the movies—came to be seen as essential for the incarcerated person's reintegration into society. Media use was seen as enabling incarcerated persons to keep in contact with "normal" society outside of prison walls and maintain citizenship and "public connection." It was also seen as maintaining the structure of a "normal day" within the prison: eight hours of work, eight hours of sleep, and eight hours of government-mandated leisure time that could be filled with media use.

Prison administrations have increasingly discussed how media technologies can be used to achieve and maintain control of incarcerated individuals, a crucial question in all punitive practice. Television proved to be an efficient tool for control, and the introduction of in-cell television from the 1970s and 1980s in Sweden has been shown to contribute to the regulation of behavior. Not only does it have a "sedating" function by keeping incarcerated individuals calm, but access—and the threat of removing access—can also be used to achieve social control (Knight, 2016). Another important dimension of prison media is the prison library (Bowden, 2002; Conrad, 2012) that research has explored in terms of its developments and circulation policies as well as the role of libraries and librarians in the rehabilitation of incarcerated individuals. In the period of the industrial prison, it also became increasingly common and encouraged by prison administration for incarcerated individuals to engage in media production themselves. A range of newspapers written, produced, and printed by incarcerated persons were launched in the 1940s, 1950s, and 1960s. The newspapers contained programs for social activities, gossip, humor, news (often related to penology), and opinion pieces (often concerning the state of Swedish prisons and work within them). Politicians, prison administrators, and other decision makers also contributed with material to these papers. Later, prison papers were followed by radio broadcasts produced by incarcerated individuals. In the case of the Kumla prison, for example, the radio station Kåkradion (Quod Radio) broadcast from the beginning of the 1970s until the mid-1990s.

Most importantly perhaps and strongly emphasized in the new penal policy from the early 1970s in Sweden, the period saw therapeutically inspired methods (besides the work in the workshops) as the main way to reform incarcerated individuals. New professions such as psychologists, therapists, and educators entered the facilities. The public investigation from 1971 (SOU, 1971:74) that preceded the policy argued that "inmates should have reasonable possibilities to pursue their personal, special interests and fulfil their needs for entertainment. They should be afforded the opportunity to participate in *study circles* and in *club activities* within the institution" (SOU, 1971:74, p. 23, English in original, emphasis added).

If the modern prison was a place of solitude where the incarcerated person was expected to listen to messages disseminated from prison priests and wardens and communicate with oneself through introspection and God through prayer, the industrial prison and postwar penology represented a regime of dialogue, interaction, communication, and community.

In the ongoing shift toward a neoliberal penal regime and prison decentralization, the communicative ideals and ideologies were once again transformed. The Swedish Prison and Probation Service was computerized during the 1970s; likewise, the technologies for mediated surveillance were developed and implemented on a greater scale. At the same time, media use by incarcerated individuals was also a hotly debated topic. During the 1950s, 1960s, and 1970s, media use among incarcerated persons had been encouraged and even mandated by law. In the 1980s, new media technologies entered the prison but not without public debate. For instance, the in-cell television system that was introduced at some of the larger prisons in the 1970s and early 1980s was perceived as being too much of a luxury for incarcerated individuals. Victoria Knight (2016) observed a similar discussion around in-cell television in the 1990s in the United Kingdom.

Public debates about incarcerated persons' access to certain media technologies continue to be controversial and spark a more general discussion about the purpose of punishment through prison sentences. Mobile phones have increasingly been considered a problem since the 1990s, when this form of contraband first became common in prisons. At the turn of the millennium, the issue of internet access became a focal point for debate: should incarcerated individuals be allowed to have access the internet, and if so how should their access be regulated? In earlier years, the large public report *Correctional Services within Institutions* (SOU, 1971) had presented television and radio as essential tools for the rehabilitation and reintegration into society of incarcerated individuals, arguing that they all should therefore have access to television and radio within the facilities. In 2005 *A Penal System for the Future*, the first large governmental report on the penal system published since 1972, was completed. The fundamental assumptions grounding this new report had changed considerably since the 1970s. Media access was no longer presented as a right: instead, it is framed as a "privilege" for well-behaved incarcerated individuals, one part of a larger "privilege system" suggested for use in the prison service. While this idea was never realized in prison administrative policy, its suggestion testifies to the total transformation of penal ideology that occurred in this thirty-year period. The positive ideal of communication as a tool for reform and the focus on human interaction and self-expression that were part of the earlier penal regime have increasingly been replaced with ideas of technology that interact with (i.e., reform and survey) incarcerated individuals and automate more and more of the interactions within prison. The penal history, then, can also be interpreted as a form of media and communications history in which communicative ideals, media technologies, and media infrastructure, together with other social and cultural changes, produce different penal regimes that crystallize and materialize in different kinds of policies in prison practices as well as in architecture and buildings. In the following chapters, we outline these changes and highlight the relevance of prisons for media infrastructure work, media as architecture, and media technologies.

3

MADE IN PRISON: PRISON MEDIA WORK

Work has been central for shifting penal regimes across history. It has ranged from the idea of hard work as a form of punishment to the idea of work as an integral part of rehabilitation and the establishment of normalcy during the prison sentence. In the novel *Grundbulten* (The foundation bolt), the prison facilities are introduced in a way that underlines this centrality of work: "In the center of this modern fortress rises a huge complex that can be seen far beyond the walls, the factory complex: giant pipes and chimneys high up in the air, carpentry, mechanical workshop, printing, cardboard production. This is the heart of the industrial prison" (Ahl, 1974, p. 85). In the mid-1970s when *Grundbulten* was written, prison work had become large-scale and industrial, and prison production was the third-largest industry in Sweden.

Part of prison work was and is dedicated to media and communication in a broad sense. Of the three categories that the Swedish Prison and Probation Service has used to organize prison work throughout the twentieth century—namely, agriculture, forestry, and industry—industrial work conducted within prisons has often included media work. Today this production has its developed own brand, "Made In Prison," to market its products. From the telegraph and the canal network to machine learning and artificial intelligence (AI), prisons and incarcerated individuals have been a crucial—but largely invisible—labor force underpinning media infrastructures and networks of communication. Nowadays prison media work takes slightly different forms. In 2019, the Finnish Criminal Sanctions Agency collaborated with the start-up company Vainu to train AI to sift through a global database of businesses to predict the best business opportunities for its clients. To generate training data for their self-learning algorithms, Vainu has initially worked with Amazon's Mechanical Turk—an online platform that matches digital workers with employers for micro work assignments—to hire digital workers to classify content such as different images. "Mechnical turkers" do what AI has so far been unable to do: they interpret context and feed the data back into the system to inform machine learning. This works well for source material that is published in English; however, there are not enough cheap workers in Amazon's Mechanical Turk who can perform contextual interpretation in Finnish and other less common languages. In addition, in the context of the Finnish welfare state it seems hard to recruit gig workers to the same extent as in the US (Lehtiniemi & Ruckenstein, 2022). As a result, the company had to find a substitute. According to Tuomas Rasila, chief technology officer at Vainu, the solution was found by coincidence. Vainu happened to share an office building with the Criminal Sanctions Agency, making it possible for them to set up a collaboration quickly. The work requires no specialized skills beyond basic computer skills, the ability to read, and the ability to interpret context and idiom in Finnish. Rasila has described the interface as game-like, involving very simple classification tasks. The incarcerated individuals do not need extensive training, and since the custom-made laptops used to complete the work were delivered to the prisons by the company, no additional equipment was required. Vainu pays its workers by delivered unit, so Raisla and his colleagues claim to know very little about the digital workers in the midlevel-security prison facility. Just like the digital workers on platforms such as Upwork and Amazon's Mechanical Turk, the incarcerated individuals working for them remain invisible. At the same time, Raisla sees outsourcing this work to prisons as a contribution to the rehabilitation of Finnish incarcerated individuals. The products Vainu sells—business predictions, advice, and connections—are based on the latest technology; as such, he has stated that he is proud to provide opportunities for incarcerated individuals to be part of such a cutting-edge technological development. After the takeover of Vainu by another company, the project was discontinued. While the company quickly moved on to new projects, the Finnish Criminal Sanctions Agency made several unsuccessful attempts keep the project (Lehtiniemi & Ruckenstein, 2022).

The kind of digital work described above fits perfectly into the contemporary prison context; at the same time, it mirrors many features of historical forms of media work and work more generally conducted by incarcerated individuals, namely work that is often low-skilled and repetitive without the need for extensive training. Prison media work has hence been mostly but not solely about constructing, repairing, and maintaining communication infrastructure or components for such infrastructure, such as the production of cables, poles, and microelectronics. Few have considered the relationship between prison work and media work because the work conducted by incarcerated individuals most often includes manual and industrial tasks in general not associated with the alleged expressivity and creativity of cultural production.

Theoretically, we can make sense of prison media work through the lens of infrastructure studies. In recent years, critical infrastructure studies have turned our attention to the social practices and work that go into developing, building, repairing, and maintaining infrastructures of communication (Parks & Starosielski, 2015). Geoffrey Bowker and Susan Leigh Star (1999) have pointed to the importance of making visible the ordinary work that is involved in infrastructure building and maintenance, including the work of janitors and cleaners (see also Parks & Starosielski, 2015). This includes studying the work that emerges with and through media infrastructures, such as Downey's study of telegraph messenger boys (Downey 2002), the media work of 9-1-1 dispatchers (Ellcessor, 2021), and the hidden work of content moderators (Roberts, 2019) and other digital gig workers (Irani, 2015). The integration of analyses of labor with questions of infrastructure points also to their sociotechnical character and the fact that technological infrastructures must rely on human productive activity. Work and labor in modern capitalist society always imply power, authority, and possibility for resistance, and these dimensions are crucial for understanding why and how infrastructures are realized and how they work. The work that goes into building, maintaining, and repairing infrastructures has often been invisible and has had a low status. Within media studies more generally, this work has only lately been integrated into the analysis of "media work." Similarly, Edwards et al. (2009) have called for more attention to "those at the receiving end of infrastructure—those who are subjected to its distribution regimes and marginalizations in everyday life" (p. 372). This includes the workers and producers of infrastructure as well as the users whose work is invisible and to a larger extent is rarely valued.

Beside continuities, prison media work has undergone important changes over time. Historically, the development of industrial prison work can be divided into roughly two periods. The first period stretches from World War II until the early 1980s, and the second period extends from the 1980s to the present. Together, these two periods point to a shift from manual production to the passive work of being tracked. These two periods also mirror the broader social, economic, and cultural developments that we discussed in the introduction, specifically the emergence of the social welfare state and its subsequent dismantling, the rise of social capitalism, and the emergence of surveillance capitalism. The first period witnessed the growth of modern state administration dominated by large bureaucratically organized groups and companies that serviced the needs of the late capitalist or "monopoly capitalist" society (Baran & Sweezy, 1966) emerging after World War II. The second period is marked by neoliberalization, globalization, and digitization, developments that all have strongly affected the organization of work in prisons. The 1980s are also marked by what David Garland (2001) has called the "punitive turn," with an exponential increase in the number of incarcerated individuals across different jurisdictions.

Prison media work follows general changes in work and labor relations related to media related value production, for example a move from manual and industrial production toward more passive forms of work that contribute to the general development of surveillance-based data capitalism as described by, for example, Tiziana Terranova (2000), Christian Fuchs (2009), and Mark Andrejevic (2007). Besides engaging in manual work, incarcerated individuals are increasingly conducting "the work of being watched" (Andrejevic, 2002); that is, they are providing and generating the data that serve as an important resource to train and feed machine learning–based technologies. Further developing Sut Jhally and Bill Livant's (1986) classic argument that audiences conduct work while watching television in exchange for payment in the form of programming content (i.e., they generate value by consuming advertisements), Mark Andrejevic (2002) argues that in the area of interactive television, this work is extended through the increased surveillance of audiences, which is what he calls the work of being watched. While Jhally and Livant argue that the productive aspect of watching television is primarily based on the acceleration of product circulation, Andrejevic suggests that online activities are valued insofar as they are seen to build niche markets through the surveillance of audiences, a process that further intensifies the cycles of capitalist production and circulation. Surveillance capitalism demands a new kind of work, the work of being tracked. The work of being tracked is more focused on data collection and analysis than previous forms of surveillance and the work of being watched. This kind of work is characterized by the value production through dispossession of intimate experience through turning them into observable, measurable units that Shoshana Zuboff (2019) calls behavior surplus. This, indeed, is a new form and shape of productive labor within (and outside) of the prison context. However, it also has a longer prehistory and shares many features with media work of the twentieth century. The work of being tracked, as with other forms of work in relation to digital platforms (as the example with the start-up company Vainu illustrates), is a form of "back-end" work, a practice that constitutes a necessary physical infrastructure for digital culture's front end that we as mundane users experience (Parks et al., forthcoming). Expanding the concept of media work to such back ends allows us to see how the captive labor of incarcerated individuals has contributed to media and communication infrastructures throughout the history of the modern prison. This can be seen as both a fulfillment and an extension of the logics of surveillance inherent and invented in the modern cell prison as such in the nineteenth century (Foucault, 1979). In the era of datafication in which more and more parts of our everyday lives are being turned into lucrative data, this work of being watched is increasing exponentially. It is also reshaped into a *work of being tracked* that takes place through digital platforms and "smart technologies." Not only is the work of being tracked that incarcerated individuals perform a key dimension of prison media work, a form of work that relates to societal transformations in labor and being (Cheney-Lippold, 2017).

FIRST WE BUILD THE FACTORY, AND THEN WE BUILD THE PRISON AROUND IT

Since the inception of the modern prison, work has been an integral part of life in prison. It was not, however, until the 1950s and 1960s that prison work became properly industrialized and large-scale factories were constructed within the prisons. Torsten Eriksson, former director general of the Swedish Prison and Probation Service, quotes the motto of the architectural committee in his book on the penal system in Sweden: "first we build the factory—and then we build the prison around it" (T. Eriksson, 1966).

Opportunities for meaningful work have been imagined as central to the modern cell prison system since its inception. This was especially true in the 1950s and 1960s in Sweden when many of the large-scale facilities built were planned according to the idea that work should be made part of incarcerated individuals' everyday lives. In the plans for new facilities and prisons in postwar Sweden, factory work was understood as an integral component for the funding of new facilities. The Swedish Prison and Probation Service's Building Committee argued that the investment in prison factories (facilities, machines, staff, etc.) should not be included in the same budget as other prison investments. The factory part was to be understood as a public commercial enterprise that refinances itself. The committee pointed, for example, to the fact that for the fiscal year 1964–1965, prison production had resulted in a surplus of 6 million kronas for the state and that prison factories, if expanded, could be even more beneficial for public authorities. The committee also argued for proper market salaries for the incarcerated individuals engaged in production, since they thought that incarcerated individuals then could "reimburse the public sector for the costs of the penal treatment" (i.e., pay for rent and food during their time in prison) (SOU, 1959). A hopeful article from 1963 in the prison paper *Hallbladet*—produced by incarcerated individuals at one of Sweden's largest prisons and a long-standing platform for critical discussions of penal policy both within and outside the prison—mentions the coming of a "million-dollar industry" that is going to be developed at the facility as soon as the new modern factories were built. The article raised hopes for not only increased productivity but also increased well-being, since it would be better and more satisfying for the incarcerated individuals to work with modern factory equipment and machines.[1] Work in prison was (and is), however, more than an economic question. As shown by Swedish prison historian Roddy Nilsson (2017), work was understood as a key element in the treatment and care of imprisoned individuals, a mechanism for resocialization and future reintegration into normal life. This also mirrored the strong moral charge given to work in Social Democratic welfare states. The social "role model" par excellence during the period after World War II was, according to Nilsson, the "loyal and conscientious working-class man who contributes to the welfare of the whole society" (Nilsson, 2017, p. 43). As a result, work became the principal treatment method in the welfare state prisons. The new prison facilities built in the postwar period were called "industrial prisons." Constructed with modern workshops and stimulating working environments, they ensured that work was at the center of prison life.

During the last 40 years, the global prison population has grown exponentially in many countries. This is especially true in the United States and other countries with partly privatized prison systems, such as the United Kingdom (Blumstein, 2011). This historical period in the global penal regime since the 1980s until today has been described as the era of mass incarceration (Wacquant, 1999). The global development of mass incarceration follows patterns of structural injustices of race, gender, and class in which policing, control, and criminalization have mainly affected marginalized groups in many parts of the world (Sudbury, 2014). The United States is by far the world leader in incarceration, with 698 imprisoned citizens per 100,000 people in 2019 (Sawyer & Wagner, 2020). This era of mass incarceration has both political-economic and ideological explanations. The end of a Keynesian paradigm in the global economy and the triumph of neoliberal economics have precipitated a global spike in unemployment rates. The prison functions as a social institution that absorbs the negative externalities, that is, the social costs of market fluctuations in the economic system.

At the same time, an ideological shift has accompanied the above economic developments and contributed to rising prison populations, as political agendas claiming to be "tough on crime" have been winning ground since the 1980s, tilting the penal debate in favor of longer sentences, the criminalization of a broader range of behavior, and incarceration as a punishment even for petty crimes. Here, the media has played a key role to "reinforce the legitimacy of mass incarceration as a remedy for social deviance" (Novek, 2009, p. 377; Yousman, 2009). The so-called war on drugs in the US in the 1980s and 1990s—with local versions in many countries—has also contributed to increased incarceration through expanding the list of illegal offenses and increasing racial profiling (Blumstein, 2011). Furthermore, these developments have been accompanied by an increased privatization of the prison system in many parts of the world. This might have had at least some effect on incarceration rates even though this is a disputed issue (see, e.g., Gilmore, 2007).

It is not, however, only private prisons that contribute to a commercialization of the prison system. Commercialization is a broader process that exists also in countries without private prisons. Except for private and commercial prisons that increase incentives for criminalizing more behaviors and growing the prison population, commercialization also includes prison labor and the capitalization of unpaid/underpaid labor (Thompson, 2012). Furthermore, commercialization engenders the use of advanced technological solutions to guard and manage the prison population to minimize the cost of salaries for human guards (Delgoda, 1980). It is in relation to these three aspects of commercialization—privatization, capitalization of prison labor, and automated surveillance—that the notion of a prison-industrial complex has been developed. The prison-industrial complex refers to an interweaving of private business and government interests, with the two-fold purpose of profit and social control, and forms the background of prison media work (Goldberg & Evans, 2009).

The industrialization of prisons and the accentuation of prison work in the postwar period were also met with fierce critique. While first having been seen as a solution to the problems of crime and "the criminal," prison industrialization and prison work were gradually reinterpreted in the 1960s and 1970s as part of the problem itself (Nilsson, 2013). Work was increasingly seen as the opposite of proper treatment and rehabilitation, as it took time and effort away from more individualized treatment programs and education that, it was argued, would be better for preparing incarcerated individuals for life outside of prison. This reorientation of the penal paradigm had its roots in academic and political discussions as well as in discussions among interns themselves. In the prison paper *Hallbladet*, this debate was lively during the 1960s and 1970s, as in this editorial from 1967:

> Our prison system is absurd since we think that we can cure the criminal with modern industrial workshops and comfortable cells. It is nothing other than "parrot thinking" [to mindlessly repeat phrases]. It is as foolish as if hospitals would stop giving medical treatment and instead practice industrial production in the hospital wards. (*Hallbladet*, 1967, Christmas issue, p. 9)

Work was also criticized on other grounds: for being too routinized and standardized, for not giving proper preparation and training for the labor market outside of the prison, and for being undercompensated with low wages. Work was understood as partaking in an ongoing infantilization of the incarcerated individuals that was not helping them to be rehabilitate, as expressed in this "letter from an incarcerated":

> In the morning the prisoner is woken from the uncanny rattling of keys. Already at this moment the daily infantilization begins. One hour after the awakening the prisoner is commanded to walk the half-kilometer through a concrete-dusty culvert low on oxygen to the "mine" = the factory. Throughout the day, the prisoner is then told what to do and how. He never has to think an independent thought, never take any initiative. (*Hallbladet*, 1972:8–10, p. 10)

Similarly and from the same time period, a series of poems written by an anonymous Swedish incarcerated person depict prison work. The poems show how the work transforms the worker, how it inhibits his thoughts and disciplines his body and mind, but without giving any hope of a future resocialization or reintegration in life and instead preparing the incarcerated for "the coming darkness."

Mechanical Potemkin Village
Monstrously disturbing
the electric wolf howls through the workshop
Muffled growls, it grabs around your throat
chews it to a vacuum of exposed nerve fibers
Deceptively soft
the welding flame caresses your iris
as a tribute
to the coming
darkness

The poem compares the prison workshop with a "Potemkin village," which also gives it a wider social significance as if the modern and rational workshops in the prisons were mainly a facade trying to convince people of the success of the modern prison as a social institution. As such, it is a good example of how the critique of prison labor—here exemplified with a voice from an incarcerated individual—was part of a more general critique of the prison as institution. This debate was particularly intense in the late 1960s and 1970s in Sweden and internationally. Around this time, several academic publications also summed up a strong critique of the modern prison system and its consequences, including *The Prison Community* (Clemmer, 1940); *The Society of Captives* (Sykes, 1958), Martinson's metastudy of rehabilitation and recidivism with the title "What Works?" (1974) and the implicit answer, "nothing" (Martinson, 1974); and Mathiesen's *The Politics of Abolition* from 1974 in which the author examines the Scandinavian prison system critically and ultimately calls for the abolition of prisons (Mathiesen, 1974, 2016). In the early 1970s, an international movement criticizing the conditions of incarceration emerged that culminated in the prison abolition movement. The movement included, among others, the Prison Information Group around Michel Foucault that was active between 1970 and 1971 (Thompson & Zurn, 2021). This group focused on the circulation of information about the inhumane conditions in French prisons. The critique of the prison system and the organization of incarcerated individuals and activists outside of prison in associations working for prison reform also spurred several prison strikes during which the incarcerated individuals refused to work and requested the right to unionization.

This critical discussion and prison abolition activism to some extent reformed the penal paradigm. Work was no longer singled out as the most important method for rehabilitation and reintegration. However, the movement did not change the fact that industrial production remained an important part of the everyday lives of most incarcerated individuals. Prison work still has an important role in modern prisons even though structural changes in the global economy as well as changes in the prison population have led to a decreasing amount of prison work.

WORKING AT THE PRISON FACTORY

Organizing work in prisons is a difficult endeavor characterized by several conflicts and ambivalences: between production efficiency and accommodation of prison routines, between support of incarcerated individuals in their rehabilitation and punishment for their offense, and between security and creativity of the work process. It is also difficult to describe the typical way in which the prison factory is organized. In Sweden as in many other jurisdictions, the ways in which the workshops and assignments are administrated and executed vary greatly between different facilities. However, there are certain experiences and aspects of prison work that are shared across the facilities.

To illustrate the character of prison work we zoom in on Kumla Prison, built in 1965 as the biggest high-security corrections facility in Sweden. Kumla is one of the so-called industrial prisons that were supposed to contribute extensively to industrial production in Sweden. Kumla is in all respects extreme in terms of size, population, employees, and workshops but also governance. While smaller facilities need support by the central administration to secure work assignments, Kumla—mainly because of its size and management—needs little centralized support. Instead, the workshops have developed their own industry contacts and prototypes for production. Around 40 employees are currently engaged in work-related operations for the 13 workshops of the facility including production managers, project leaders, and workshop heads. Kumla houses around 430 incarcerated individuals who are living in different housing units and strictly separated from each other. One workshop always belongs to the same housing unit, and incarcerated individuals cannot change between different workshops. Hence, incarcerated individuals' work assignments rarely match their skills or interests and mainly depend on their assigned housing unit. This reflects a problematic tension between the needs of assigning housing units based on security measures (including sentence length and potential conflicts with other incarcerated individuals) on the one hand and work-related needs on the other. At the same time, many incarcerated individuals have difficulties focusing on work over longer periods of time, and work operations are often merely seen to be used to pass time rather than produce certain units in an efficient and rational manner. Additionally, working days are shortened by therapy sessions and mere movement between the different blocks that are connected through an elaborate culvert system in this extensive facility. While a working day is supposed to last for six hours, an incarcerated person effectively works for four hours a day on average. The effective working time can, however, range between five minutes a day to the full six hours depending on the preconditions of the incarcerated person. The work morale at Kumla seems high in comparison to other facilities. One of our informants—the manager of work operations at Kumla Prison—told us a story from another Swedish prison where he worked for a shorter period. One of his colleagues called him and frenetically screamed "You have to come down to the workshop and see this. You won't believe it, something has happened. An inmate is working, can you believe it??? An inmate is working!"

MEDIA WORK IN THE INDUSTRIAL PRISON

The Kumla prison was one of two facilities that was planned during the 1950s and opened in the 1960s and was supposed to replace an older prison, Långholmen in Stockholm. In this older prison, the main work operation had been printing:

> When I started working in the printing workshop right before the Second World War, the incarcerated individuals had long sentences and could really learn the craft. The printing workshop went well, and it was a good atmosphere. I attended the supervision training but didn't have to attend to much surveillance. There were others who could take care of that. In our group, we just did our job. With the criminals during Prohibition, many fine professionals arrived at the

> printing workshop and there were a lot of nice people I got to work with. It was only once I met a prisoner who was a little hot-headed. In the beginning, we printed the yearbook of the forensic psychiatric clinic and the employee directories. Later we printed the forms of the prison services, customs and the forest services. We also printed the prison papers *Murbräckan* and *Cellstoff*, which were set manually. We produced the pages in one of the big old high-speed presses and it was exciting when the rollers went over and the press worked. If you didn't work on the letters properly or worked too fast, they jumped out and we got a hailstorm of letters all over us. I remember that we were almost terrified when we were supposed to work with these presses. Normally, we were around eleven or twelve people working in the print shop. Later on, when the prison was overcrowded, we would get in so many people that it was hard to find work for all of them and keep everything under control. We were supposed to count incarcerated individuals several times a day. When they came in in the morning, when they returned from their walk and when the working day was over. But you wouldn't point at every prisoner, but instead count a little more subtly. At the end of the working day, the prisoners were supposed to stand in line behind a fence and walk over the yard under the supervision of the guards. Sometimes there were searches if they suspected contraband. Once one person was missing. We found him later in the closet where we used to store the printing rollers. He was lying there printing. So, we had a lot of juggleries, smuggling and so on, but it wasn't that bad. (Lamroth, 1990)

This statement from the foreman who in the 1950s oversaw the printing workshop at Långholmen captures the mundanity of media work conducted in the prison as well as the ways in which the unique constraints of the prison context set the tone for the work conducted. The efforts to maintain professionalism amid the challenges posed by a lack of relevant skills and education among incarcerated individuals and those leaving at the end of their sentences and the additional task of controlling and surveying incarcerated individuals sound both different from and like what we know as media work.

The understanding of media work as something taking place within the media industries and of the media industries as companies and organizations involved in various forms of expressive-aesthetic, cultural, and informational production, dominates the research field (see, e.g., Deuze, 2007; Hesmondhalgh & Baker, 2011). Creative processes, individual intentions, and culturally and socially shaped contexts of production have hence been given much attention, somewhat shadowing the broader network of production involved in media and communication, what we might think of as the back ends of media and cultural production (Kaun et al., forthcoming). There is a broad range of such sites and practices that constitute a necessary but largely invisible infrastructure for media and cultural production. A range of routine tasks that rest on cheap (or free) labor must be performed and aggregated to make possible the front end, that is, the experiences of media user's and consumers of cultural products. These back ends also put constraints on and facilitate possibilities and frameworks for the expressive-aesthetic and cultural production normally defined as media work (see, e.g., Deuze, 2007; Hesmondhalgh & Baker, 2011). Thus, to fully understand the complex processes of media production, the full range of media work, we argue that one also must take an interest in the margins of media industry of which the prison industry is one. In that sense, the manual, routine, and piecework production of prison industries can be seen as forms of media work, and this media work has shifted in relation to more general social processes and transformations within media industries. In the industrial prisons of the 1960s and 1970s, the media work was mainly done in relation to the large, centralized, and often public media and communication companies of the time and was in general directed toward producing pieces of communication infrastructure and hardware for media industries.

This mirrors the general focus of the modern industrial prison on developing facilities for large-scale and industrial work that had not been possible in the older cell prison facilities built in the nineteenth century. The government's proposition from 1957, when the decision to build Kumla Prison was made, describes the plan for the facility as such:

> The main idea has been to, within a large institutional area, create several relatively small institutional buildings, generally one-story buildings, in such a way that large outdoor areas are created. The clientele is differentiated into groups with limited size. Work operations are set up according to factory lines.[2]

On a very detailed level, the proposition continues to describe the outline of the new industrial prison in Kumla:

> The workshops for the incarcerated individuals are intended for mechanical industry, wood industry, clothing industry and occupational therapy and includes approx. 6,700 sqm gross area including cold storage. . . . The whole factory complex forms a closed unit, thus facilitating surveillance.
>
> The mechanical workshop is set up according to the following plan. From a loading dock on the short side of the factory leads a cable car to the warehouses

> and processing premises. To the right of these are staff rooms as well as changing rooms, laundry and rest rooms. Then follows a large machine and assembly hall, facilities for surface treatment, inspection and control and a warehouse for finished items. . . . With regard to the operation of the institution, the Committee states that employment of the incarcerated individuals should be as varied as possible, since the facility will accommodate interns among which there are great differences not only in ability and willingness to work, but also in terms of interests and orientation.[3]

The workshops at Kumla Prison, with their focus on mechanical industry and woodwork, hence shifted the nature of manual prison media work from printing to infrastructural (hardware) work.

In the postwar period, large state and private companies and government administrations were the primary customers for prison-produced goods. That prison industries should have priority in public procurement processes was formal policy at the time. Expanding educational and bureaucratic organizations after World War II needed infrastructure to function, and incarcerated individuals provided the cheap labor required to manufacture the large quantities of bookshelves, filing cabinets, desks, chairs, folders, binders, and records needed. For example, Swedish incarcerated individuals met the entire Swedish government administration's need for bookshelves, delivering approximately 5,000 shelves every year. Figure 3.1 depicts one example of a shelve system that was designed and produced at Kumla for a bomb shelter beneath Stockholm University.

It might be far-fetched to consider this production as a form of media work given that the clients were not media companies. However, the incarcerated individuals produced the hardware and infrastructure required to archive, distribute, and communicate information including filing systems and shelves for the Swedish welfare state. Craig Robertson (2021) has identified such systems as being critical to the expansion of modernity and crucial information infrastructure of modern bureaucracy, capitalism, and surveillance. However, among the prisons' clients during this period, we also find several media and communication companies proper and a range of production that is more straightforwardly a form of media work. The media and communications industries in Sweden during this period were dominated by large state-owned companies such as Televerket (the state-owned telecommunications cooperation), the Royal Mail (the state-owned postal service), and Teracom (a terrestrial broadcast transmission service

Figure 3.1
The prison factory and incarcerated individuals involved in industrial production at Kumla Prison in Sweden, depicted during the opening year of 1965. Pictures originally published in the newspaper *Örebro Kuriren*, February 12, 1965. Now licensed under Creative Commons, Public Domain 1.0. Available at https://digitaltmuseum.org/

company). Among the prisons' regular customers were also semigovernmental organizations such as the public service companies Swedish Radio and Swedish Television. Communication, information transfer, and the effective coordination of "monopoly capitalism" (Baran & Sweezy, 1966) required communication infrastructure, so the emergence of modern large-scale administration and bureaucratically organized companies during the postwar period made ever greater demands on communication infrastructure of this kind (mail, telephony, broadcasting). As we can see from the table (table 3.1), work for the public telecommunication and postal services stood for between 6 and 12 percent of the total production within the Swedish prison system in the first half of the 1970s. However, large private companies such as the Swedish Film Industry and Ericsson were also among the clients ordering components for media and communication infrastructure from the prison industry. The total share of prison work that was media-related was therefore higher.

Table 3.1
Overview over parts of prison media work 1970–1974

Year	Total orders Total	Televerket	Royal Mail	% of total post and telecommunications
1970	2,150	211	52	12
1971	1,800	103	50	12
1972	1,700	72	42	7
1973	1,700	79	37	7
1974	1,700	75	22	6

Data produced by the authors based on collection: Kriminalvårdsstyrelsen Arbets- och utbildningsverksamheten_Anstalts- och kundregister_1970–1975, C1b1–6.

For the postal service, Swedish incarcerated individuals produced information boards, letter bags for postmen, letter trays, parts of postbags, air mail bags, parts for typewriters, wallets, package trays on wheels, folders, and stamps as well as pallets. Televerket ordered bookshelves, letter bags, mailbox stands, trays for catalog holders, spacer plates, spacer tubes for catalog holders, flags, parts of headphones and other small electronic parts on a large-scale every year in the 1970s. Among the list of the 10 largest buyers from Swedish prisons in the 1970s, 2 media and communication companies make the list: Televerket and the Royal Mail (table 3.2).

Clearly, media-related work has had great importance for the Swedish Prison and Probation Service, as media and communication companies during the twentieth century and up to today have been a substantial part of its customers. What has been the impact and role of prison labor for the media industries in general? The production of media and communication technologies and infrastructural hardware (cables, poles, etc.) makes up a rather small part of the overall production within these companies. The number of incarcerated individuals at Swedish prisons in the 1970s was about 12,000 per year, and at maximum a couple of hundred to 1,000 of the incarcerated individuals were involved in the production of goods for the media and communication industries (no exact figures exist, however). This can be compared to the fact that Televerket alone had 40,000 employees in 1970, the Royal Mail had 58,000 employees in 1976 (Prop. 1970:128; Prop. 1975/76:167), and the private telecommunications enterprise Ericsson had 71,070 employees in 1976 (Ericsson, 1976). For fiscal year 1975 the total production value in the Swedish prison system was 90 million kronas

Table 3.2
Largest clients of the Prison and Probation Service in 1976

Company Name	Commodity	Production Value
Nyckelhus AB	Ready-built houses	20,900,000
The Swedish Military	Equipment	7,850,000
Royal Mail	Equipment	6,400,000
IKEA	Furniture	4,000,000
Road Administration	Road signs	3,000,000
Gothenburg Municipality	Laundry service	2,900,000
Televerket (public enterprise for telecommunications)	Equipment	2,790,000
Trästandard	Furniture	1,800,000
Public agencies	Furniture	1,100,000
KF	Furniture	1,000,000

Source: *Expressen*, November 6, 1976.

(around 500 million in today's value), and the media-related part of that revenue was around 10 percent; compared to the economic turnover in the media and communication companies at the time, this was of course marginal. Ericsson, for example, had turnover on 7,312 million kronas in 1976.

The overall role of prison labor for the media and communication industries is as hard to fully assess. However, prison labor has significance for specific products that require low-skilled routine work at low costs. Torsten Eriksson emphasized repeatedly that prisons were the third-largest industry in Sweden in the 1960s. Even if the Swedish Prison and Probation Service itself states that it never competed in pricing and even though it is prohibited by Swedish law to compete with commercial companies through low prices, the pricing issue and the question of unfair competition were constantly raised in relation to prison labor throughout the twentieth century. Wages in prison labor are low; in the 1970s incarcerated individuals were paid 4–4.5 kronas per hour (wages for comparable work outside of prison earned was about 25–30 kronas per hour at the time). The possibility of cheap production was and is a main driver of the institutional couplings between the media and communication industries and the prison and probation services. Another key aspect are the general relations between large bureaucratic (often) public media and communication companies and the state as such during the postwar period. The corporatist spirit of the time forged several entanglements both institutional and individual, between large enterprise and state bureaucrats, which also made it seem natural that state-sponsored incarcerated workers should be used to support the media and communication industries.

The period between 1960 and 1980 was the golden age of industrial prisons, a period in which the prison industry was one of the largest branches in the country (see table 3.3). This era is marked by tight connections between large centralized public media and communication companies and prison production. Components for communication infrastructure and hardware were the products that the prison industry provided to the media industries. The media work in prisons has always been infrastructural work, which is true also for the periods preceding the industrial prisons and after the 1960s and 1970s. As the political economy of society and of the media industries transformed in the 1970s and 1980s, so did prison work, but it remained a form of industrial back-end–production, also within relation to the digital and increasingly globalized economy and media society of the postindustrial twenty-first century.

MEDIA WORK IN THE POSTINDUSTRIAL PRISON

During the 1970s, the nature of work within prisons changed. These changes were to a large part driven by general changes in the labor market and in society. The increasing globalization of economic life in the 1970s brought about a general structural transformation of the Swedish economy. Both inside and outside the prison walls, the manufacturing industry was severely affected by this transformation process, which implied among other things that a series of mainly routine manufacturing tasks were outsourced to low-wage countries, especially in Asia (Kindgren & Littman, 2015). Many of the customers who had previously relied on prisons for manufacturing work also turned to this new labor market. At the same time, the large state-owned companies that had been major customers in the work of the correctional services were dismantled. These changes in the economic life of society, both of which are part of the transition to a neoliberal economic-political regime, made it increasingly difficult to find customers for the prison industry and thus work for the incarcerated individuals. Another equally important factor for the decline of manual prison work in this period was the growth of so-called protected labor (Riksrevisionsverket, 1979) at the Swedish labor market. To facilitate employment for people with disabilities and those who suffered from occupational injuries or other circumstances hindering their full participation in the labor market, county councils and municipalities began to organize protected labor that ensured employment for those groups. In the 1970s, these efforts intensified and were increasingly centralized until a national public institute, Stiftelsen samhällsföretagen, was formed in 1980. Later this operation was corporatized, and it is now a publicly owned company and as such one of Sweden's largest employers, with more than 24,000 employees and an annual turnover on more than 9 billion kronas. This form of protected labor was (and still is) a direct competitor to prison work, as it represents a similar form of simple manual production competing over the same customers. As the labor market changed, the public-protected labor could more easily shift to service production (cleaning, park and gardening work, etc.), for obvious reasons this was more difficult for prison work. In the report *Arbetsdriftens vägval* (The Road Ahead for Work in Prisons) published in 1995, the Swedish Prison and Probation Service concluded that work in prisons "need to catch up with the structural changes on the Swedish labor market" and be more "oriented towards the service industry" (Kriminalvårdsstyrelsen, 1995, np). However, how this could be done was not answered in the report, and it is obvious that the Prison and Probation Service was somewhat short of answers and ideas at the time.

Following the large public investigation of correctional services prompted by the Swedish government that took place in the early 1970s, several improvements in line with the demands from incarcerated persons were also proposed, including greater opportunities for education and for other occupations than industrial work to be made available to incarcerated individuals. Already in 1968, the parliamentary committee for institutional treatment had suggested that work within prisons should be abandoned or at least no longer be mandatory. The committee discussed the question of market-based salaries and the idea of maintaining prison work only at some specialized institutions where "highly effective workshops" could be maintained and where "a clientele could be placed who could cope with a high pace of work and who do not need any other treatment than such work."[4] The background for the latter suggestion was the weak productivity that for long had been observed but also the increasing problem with drugs inside the prisons, a problem that left a larger share of the prison population unfit for productive labor. The 1974 correctional reform introduced the concept of

Table 3.3
Employment program: Hourly earnings in Swedish kronas for fiscal years 1976–1977 until 1980–1981

Year	Mechanical workshops	Agriculture
1976–1977	377	282
1977–1978	422	334
1978–1979	451	359
1979–1980	503	413
1980–1981	542	483

Source: 1981data compiled by the authors based on annual reports.

rehabilitation to the correctional service and established leisure-time services in the institutions. At the same time, decentralization took place: so-called local prisons were created to allow incarcerated individuals to work close to the local communities in which they were to be integrated after their release. At the same time, the professionalization of prison production—with a stronger focus on the work environment, occupational safety, and market wages—meant that the competitive advantage of the low wages was lost. Ironically, the efforts to improve prison working conditions reduced the demand for incarcerated individuals' work and limited opportunities for the institutions to offer meaningful workshop employment to incarcerated individuals (table 3.3).

During the 1980s, the prison administrations reported constant problems with work operations due to the deterioration of the market for prison labor as outlined above. Even as the central authority, the Swedish Prison and Probation Service, demanded better profitability and more customers, the demand for incarcerated individuals' work fell. The institutions' annual reports from the 1980s return again and again to the problems with prison production. The institutional directors also used these annual reports to criticize the management and central authority on this issue. For example, in the annual report of Hällby Prison for fiscal year 1987–1988, the director wrote the following:

> The past year has largely gone without disturbances, but some problems have proved to be in full bloom. It is primarily the workshop operation that has had difficulties keeping the production going, partly because several supervisors have ended their employment and the appointment of new supervisors has gone very slowly.

> Immediate action is needed to correct this situation. We cannot accept that the central authority imposes increased demands on production and at the same time does nothing to enable increased production to become a reality. If we are to seriously believe in all the talk about the importance of a balanced budget, we must have concrete proof that the authority is serious about giving us the opportunity to achieve a zero-result here on the spot. I think the institution is well on its way to getting much better on the workshop side, but that a tangible indifference to the workshop problems can cause the workshop staff to become tired of constantly going that extra mile without getting either thanks or support from the central authority. (p. 1)

One problem that also loomed large in this decade was low productivity, which came at least in part from recurrent work refusal (a growing phenomenon that was called "passivity in the workshops"). This problem has characterized production throughout the history of the modern prison: even in the 1920s and 1930s, the reports of the old Royal Prison and Probation Authority complain about a lack of discipline and work ethic in incarcerated individuals. Further, because more and more incarcerated individuals were very young and sentenced to rather short penalties, the prison population was constantly being replaced, and it was hence difficult to retain trained staff in production.

DISPLAY AND DISTRIBUTION

Although the 1970s and 1980s saw a decline in production, prison work obviously did not stop altogether. While fewer goods were ordered from the Swedish Prison and Probation Service, clients were still present. As of 2018 there were at least 37 different media companies in the customer register of the Swedish prison production (table 3.4). It is still common for prison interiors and prison clothes used as costumes for TV and film productions to be made by incarcerated individuals in prison workshops. The product catalog from KrimProd, which lists their own products (not those subcontracted and produced on demand) include media and communication artifacts such as TV boxes, newspaper stands, acoustic panels, street stands, computer shells, envelopes, and correspondence cards.

The customer register naturally only tells the story of which companies have placed orders with the Swedish Prison and Probation Service and nothing about what kind of products they have commissioned. In interviews with the production managers at the Kumla and Hall facilities, products

Table 3.4
Media and communication companies in the customer register of the Swedish Prison and Probation Service in 2018

Company name	Area of business
Aftonbladet	Journalism
Brand New Television AB	Media production
Canon	Electronics
Dagens Nyheter	Journalism
Duplica Print & Kommunikation	Advertising
EO Grafiska	Graphic design
Eskilstuna-Kuriren	Journalism
Esselte	Graphic design
Exakta Printing AB	Advertising
Expressen	Journalism
Göteborgsposten	Journalism
Svenska Dagbladet	Journalism
Helena Ullstrand Design AB	Advertising
Helsingborgs Dagblad	Journalism
Hexatronic Fiberoptic AB	Electronics
I-magé of pr AB	Advertising
Ink n Art Produktionsbyrå AB	Graphic design
Mittmedia AB	Media production
Nordisk Film TV-Produktion AB	Media production
Nyhetsbyrån Siren	Journalism
Piteå-Tidningen	Journalism
Print on demand Stockholm AB	Graphic design
Profilskaparen	Graphic design
Promotion i Boo	Advertising
R.M. Electronic AB	Electronics
Realtryck AB	Graphic design
Ricoh Sverige AB	Electronics
Rodemreklam AB	Advertising
Sib-Tryck Holding AB	Advertising
StrieProfil AB	Advertising
Sveriges Radio	Media production

Table 3.4 (continued)

Company name	Area of business
Sveriges Television	Media production
Sydsvenskan	Journalism
Sörmlands Media AB	Media production
Tele2 Business AB	Telecom
Teq Display	Advertising
Tredje Statsmakten Media AB	Media production

for display and distribution, often to the advertising industry and for media forms such as billboards, signs, and display stands, are pointed out as major products delivered by the prisons. Media work in prisons has hence gone from gravitating toward hardware for communication infrastructure and larger (public) technical systems for distribution to becoming more aimed at infrastructures for advertising media and media display.

This form of work highlights a tension that runs through the cultural history of the prison as such, between invisibility and visibility. As discussed by the Swedish historian Dan Waldetoft (2005), the rise of the modern prison in the late nineteenth century meant that incarcerated individuals were taken out of public view. Prior to modern prisons emerging in the late nineteenth century, incarcerated individuals often were seen working outside of the prison facilities, and executions as well as transports of convicts were often public events in which the incarcerated individuals were made visible as a spectacle for citizens. This changed toward the end of the nineteenth century (the last public execution in Sweden was held in 1877), and during the twentieth century incarcerated individuals were increasingly hidden from view, and prisons became secluded and "dark" and "mysterious" places (Waldetoft, 2005). In the cell prisons of the late nineteenth and early twentieth centuries, incarcerated individuals were not even visible to each other; they were separated by screens or covered by masks when they were in the communal areas of the prisons (which happened rarely) and were otherwise isolated from each other in separate cells. However, the cell prison as such was typically placed visibly, often in city centers. Moreover, the seclusion and invisibility of the people within the prison opened new forms of visibility: on the one hand the surveillance in the prison where the incarcerated individual was always subject to a surveying gaze and on the other hand the media spectacle or media fantasies of the secret world of the prison as imagined in popular culture. A similar development in media work is the fact that when incarcerated individuals were more visible to the public eye (e.g., working outside of the prison walls), they produced mainly "invisible" media infrastructures such as digging for underground cables. The media work by the invisible incarcerated person of the contemporary prison is, on the contrary, highly visible in society and is often technologies for producing visibility in so-called display media. One example of such a product is the newsstand used to display newspapers and magazines, which is common in almost all grocery stores and kiosks in Sweden. Such stands are produced in the factory at Kumla Prison (figure 3.2). In this way, the largely "invisible" work of incarcerated individuals is in fact highly visible in public settings all around the country.

The newsstand also underlines another long-standing tradition within prison media work: the relation between newspapers and incarcerated labor. The history of newspapers and print journalism has mainly focused on journalists, newspapers as organizations, journalistic content, and the role of print journalism in society (see, e.g., Wahl-Jorgensen & Hanitzsch, 2009). Less acknowledged in these histories is the processes and work that make newspapers as such possible from the outset, such as printing and distribution. In both areas, incarcerated individuals and incarcerated labor have a role throughout media history.

Printing technologies were vastly improved during the nineteenth century, making mass circulation of newspapers possible. Even though automation was increased, by the end of the century printing was still a craft that needed considerable amounts of manpower (Musson, 1974). Newspapers generally had their own printing operations and needed (cheap) labor to operate them. For this purpose, people otherwise excluded from the labor market were used to staff the print shops, such as persons cognitive impairment. Incarcerated individuals were also a category of workers often hired to operate the printing of the new mass press of the late nineteenth century (Kellokumpu, 1997, p. 149). This was at a time when incarcerated individuals could work outside of prison facilities, return after a working day, and hence could be hired by newspapers. Later printing operations moved into the prisons themselves, but in these print shops newspapers were rarely printed, and incarcerated labor had no role in printing newspapers during the twentieth century.

In the industrial prisons developing in the mid-twentieth century, newspaper distribution and circulation instead became part of the work of

Figure 3.2
Newsstand produced in Kumla in a grocery store. Photo taken by the authors.

incarcerated individuals. Naturally, incarcerated individuals at the prisons were not directly involved in distributing newspapers. However, companies for newspaper distribution from the 1960s and onward had been important customers of the Swedish Prison and Probation Service. In Sweden, carriers have been responsible for distributing newspapers to subscribers' homes during nighttime and early morning, and delivery has been a service provided by the newspapers themselves (Engblom, 2020). Newspaper carriers have generally been employed by the newspapers directly, but there has also been a parallel system with self-employed carriers working on commission. The profession started to develop in the late nineteenth and early twentieth centuries and was at the outset a gig economy dominated by dayworkers with very low, if any, income: the "wage" for carrying newspapers at the time could be a pair of shoes, for example (Kellokumpu, 1997). Unionization among newspaper carriers took off quite early, and already in 1905 the newspaper carriers formed a union. This was spectacularly early, not least since newspaper distribution at the time was a form of work entirely dominated by women, and women-dominated sectors were generally not as quick to unionize (Wrenby, 1955). The union struggles included many dimensions, but one aspect with relevance here is the demand for better working equipment and appropriate aids and tools for newspaper distribution. In documents from the recurring contract negotiations between newspaper carriers and the employers, protective clothing, carriages, shoes, gloves, and so on were issues repeatedly on the table in negotiations for national agreements.[5] The printing ink destroyed the carriers' clothes and penetrated their skin. The increasing thickness of newspapers led to an increased weight from around 33 grams per issue at the turn of 1900 to as much as 500 grams toward the end of the century (Kellokumpu, 1997). Such matters called for better working equipment. The demands for better working equipment were slowly met in negotiations between the unions and the employers during the twentieth century. Newspaper distribution is, however, an area in which the news organization wants to keep costs to a minimum, and this explains why much production of working equipment for newspaper carriers has been conducted in Swedish prisons.

Bags for carrying newspapers were commissioned from several Swedish penal institutions and produced in large quantities. At Kumla Prison, a trolley (figure 3.3) was at first commissioned and later became one of the facilities' licensed products, as they overtook the rights and patents for the

Figure 3.3
Trolley for newspaper distribution, 1980s.

construction. In this way, prison labor produced parts for the "soft infrastructure" (Forsler, 2020) of newspaper distribution, the network of invisible carriers that during the dark hours of the day moves the news from the printing shops to the homes of individual subscribers. The carriers of newspapers upholding the distribution infrastructure had a major role in deciding the size and weight of newspapers. This invisible back end of the newspaper industry hence was important for how much news and informational material each issue of a given newspaper could contain. To allow for heavier newspapers with more pages and more material (and more advertising), the newspaper industry turned to the Swedish prison industry, which developed trolleys and other equipment for carriers and produced this equipment cost-efficiently enough for it to be profitable. Hence, a series of processes in the back ends of media production and distribution had a decisive impact on and role in forming and guiding the editorial decisions on what to include in a newspaper.

In the mid-1970s and early 1980s, there was a downturn in prison production more generally. The large centralized public media and communications industries that had been major customers were dissolved or privatized, an their production was outsourced abroad. There were clear difficulties in switching production and adapting to a transformed labor market. Still, industrial and manual production of media infrastructure and hardware remained an important part of prison work (and is still today). As in the earlier periods and in the period that would follow, it was the backend production of technologies and hardware for the media industries that was the role of prison work for the media. In this period, media for display and distribution became important parts of the media work in Swedish prisons, also mirroring more general shifts in the media industries, now more based on advertising and increasingly dispersed and decentralized networks for distribution. In the period to come, digitalization and the ideas of the "smart prison" increasingly came to affect media work in the prison context.

MEDIA WORK IN THE SMART PRISON

Prison media work has been characterized by several changes since the postwar period. Moving from manual work in the 1940s to a steep decline of prison media work during the 1970s and 1980s and a reemergence of that work in the 1980s, current prison media work is characterized largely by the principles of digitalization. Here the notion of the smart prison—a prison reliant on digital devices for the sake of increased reliability and efficiency—is key for understanding the latest articulation of prison media work, namely the work of being tracked (and consequently datafied). In the Swedish context, these developments are palpable in the approaches toward digitalization promoted by the Swedish Prison and Probation Service. In 2018, the state agency launched the initiative Krim:Tech in order to recruit technology developers. Additionally, Krim:Tech is meant to serve as a hub to renew, digitize, and smartify the work with incarcerated individuals. In a job posting in March 2018, the Swedish Prison and Probation Service described the hub as follows:

> Krim:Tech is the new digitization initiative by the Swedish Prison and Probation Service. With the help of the latest technology and research, the initiative will support the development of new and improved digital solutions within the authority.
>
> Krim:Tech is an inventor's workshop and testbed for digital technology. Does an ankle monitor have to actually be an ankle monitor or could it be something else instead? How can we use IT to keep our security class 1 facilities calm? How can we prevent children and families who are visiting their father or mother in prison from feeling afraid? Can we do the security scan with a toy instead of metal detectors and full body scanners?[6]

Although the use of digital media technologies in the prison context is still limited, digital technologies are employed in more and more areas. The most common way that digital technologies enter prisons is as solutions offered by different companies that allow for ubiquitous computing to track, manage, and control prison populations. One example of smart digital technology in the prison context is Spartan by Guardian RFID (figure 3.4). The

Figure 3.4
Spartan by Guardian RFID. *Source*: http://www.correctionsforum.net/article/spartan-by-guardian-rfid-42511.

android-based handheld device is equipped with radio frequency identification (RFID), wi-fi, and push-to-talk as well as high-resolution imagining to be used to automate security rounds and for headcounts, activity tracking, and more. The website advertising the device encourages clients to "centralize your inmate identification, security rounds, and activity logging into one powerful platform that integrates with your jail or offender management system. Maximize your defensibility, mitigate risk, and gain lightning-fast, real-time reporting with corrections most powerful Command & Control platform."

Not only does the Guardian RFID website gather various case trials and blog posts about the advantages of moving toward RFID logging and cloud services, it also collects the voices of administrators, wardens, and officials in sheriffs' offices. Most of the blog posts and user testimonies are concerned with the increased efficiency of controlling and supervising incarcerated individuals in real time and with fewer errors. The following represents one example:

> Real-time Insight:
>
> GUARDIAN RFID helps us make data-driven assessments about inmate observation and classification levels. We're measuring staff performance on our security rounds in real-time, which helps us manage compliance with jail standards.
>
> Lt. Belinda Jackson
> Brazos Co. Sheriff's Office
> Bryan, Texas[7]

As this testimonial makes clear, Guardian RFID not only supervises and controls incarcerated individuals but also monitors the productivity of guards and other staff members. Besides selling specific devices, Guardian RFID also offers help with the complete outsourcing of server capacity and storage and processing of the data that are collected automatically.[8] By standardizing the digital technologies used to survey and control prison populations, a range of different data can be collected and stored in common databases. This storage opens the possibility for analyzing these data in real time with the aim of predicting and preempting unwanted behavior among incarcerated individuals. The Offender Management System developed by GTL, for example, promises to collect and handle "information on *all aspects* of an inmate's incarceration."[9] This would allow a full picture of the incarcerated individuals' movements through the prison and of previous records but also health measures extending and standardizing the tracking beyond previous possibilities.

Large-scale data collection and the construction of comprehensive databases is also at the center of a news story that unfolded in January 2019 when the adversarial online news publication *The Intercept* reported in an extensive article about a voice surveillance system that had been rolled out across the US. According to the article, incarcerated individuals in an increasing number of correction facilities have been pressured into recording their voices. Often without being fully informed about the purpose of the recording or given no choice but to comply or lose the privilege to make calls, a large number of incarcerated individuals have been voice fingerprinted. Voice fingerprinting works through extracting and digitizing voices of prison population and adding them to a large database. The aim is to preempt future crimes and detect criminal networks coordinated from correction facilities. The surveillance extends beyond prison gates and includes external callers. The total number of all incarcerated individuals' voices recorded and registered is difficult to estimate, but the New York Department of Corrections confirmed that about 92 percent of its total 50,000 incarcerated individuals have been enrolled in the voice-recognition system. Other states with large prison populations such as Florida, Texas, and Arkansas have confirmed the purchase of the recognition software by Investigator Pro as well. Producing large-scale searchable databases of voice fingerprints, the company Securus that offers the voice finger-printing software argues that it monitors over 1.2 million incarcerated individuals in over 3,400 facilities, but the total number of incarcerated individuals registered in that way is hard to estimate. These databases not only serve the surveillance of criminal networks and undergird predictive policing but, in a broader sense, also feed into the development of voice-recognition technology that has become a million-dollar industry (Turow, 2021). While Joseph Turow is particularly interested in the marketing strategies that are increasingly based on emotion recognition and are predictive with the help of automated analyses of voice recordings, the neural networks that allow for this kind of predictive work need to be trained. The databases emerging in US prisons are the perfect training sets.

FROM THE INDUSTRIAL PRISON TO SURVEILLANCE CAPITALISM

Regarding surveillance and discipline, Foucault (1979) argues that "the Panopticon . . . has a role of amplification; although it arranges power, although it is intended to make it more economic and effective, it does so not for power itself . . . ; its aim is to strengthen the social forces—to increase production, to develop the economy . . . to increase and multiply" (p. 208).

As a metaphor, the panopticon stands for the nineteenth-century shift toward establishing self-monitoring and self-discipline to increase economic productivity under capitalism. Within the panopticon, citizens discipline themselves into productive laborers to produce surplus value. According to Andrejevic (2007), the panopticon becomes a metaphor to make sense of emerging forms of disciplining the self and surveillance. In contrast, we turn to prison media work to trace changes in infrastructural work over time (Cheney-Lippold, 2017).

The productive labor of incarcerated individuals constructing the databases through being tracked and datafied can be understood as one form of "behavioral surplus" (Zuboff, 2019). Zuboff argues more generally that we produce value by moving through digital worlds and leaving data traces. These data are used to further rationalize and speed up production processes and to automate production through AI and machine learning. Machine learning and AI require large amounts of data to train their algorithms, and as Cheney-Lippold (2017) argues, we supply these data by living our digital lives. Incarcerated individuals contribute in several ways to the work of being tracked. Their lives are datafied unto the last detail. Records are kept about their food intake, emotional status, and media consumption. A second way in which incarcerated individuals produce value by being tracked is by serving as test subjects for surveillance technologies in a way that would encounter public pushback due to privacy concerns if it was tested out in other contexts. One example is AI-powered broad-range CCTV coverage. Footage is automatically analyzed for unusual movements and behavior, notifying guards when deviations from normal activities are detected. In Malaysia all prisons are completely covered with CCTV, but surveillance is concentrated in only one control room. Training AI in this context hones algorithms that can then be deployed in other commercial contexts. It also normalizes new and more data-driven surveillance techniques for broader use in civil society. The "work of being watched" (Andrejevic, 2002) that incarcerated individuals perform is a key dimension of prison media work; however, it is also increasingly driving broader societal transformations of how labor is structured, including the media work performed by prison guards. Much of guard work is about controlling, moderating, suppressing, or steering communication and media practices (who talks to whom, what media content can be consumed by whom, etc.). According to Zuboff (2019), under the new regime of surveillance capitalism our behavior is turned into observable and measurable units that can be processed with computational methods. She argues along several others, including Tiziana Terranova (2000) and Christian Fuchs (2014), that human behavior that is turned into data is the new raw material for surplus production. In that sense behavior, in the case of voice-fingerprint speech, is producing value in the form of data. As with previous forms of prison media work, current forms of work in prisons are intimately linked with general developments of capitalism. In the case of surveillance capitalism, however, the idea and practices—namely the surveillance, tracking, control, and prediction of behavior—that have become central to value production share many characteristics with prisons.

So far, we have traced the contours of the media work that was carried out by incarcerated individuals in Swedish prisons during the twentieth and twenty-first centuries. As we have pointed out, incarcerated individuals have contributed to many of the media and communication infrastructures that we use in everyday life today. This kind of media work differs from much of the media work and media production that so far has been at the center of media research, which has been primarily interested in the symbol-producing and expressive media work conducted by society's "creative classes." Focusing on incarcerated individuals' media work illuminates three key insights for media research. First, prison work entails a dimension that exists in all wage labor but stands out particularly clear here: work is in some ways and to some degree always unfree. It can be a useful reminder that media work, often described as "cool, creative and egalitarian" (Gill, 2002), is embedded in hierarchical power relations and inequalities in different ways. The work conducted in prisons is linked to how the labor market and the economy are largely organized in society. The development that has taken place, from an emphasis on industrial production to "passive" media work where the fact that incarcerated individuals are technologically monitored becomes a kind of work, also illuminates trends and tendencies that exist in society in general within contemporary surveillance capitalism. Second, considering prison work as a form of media work deepens our understanding of the political-economic contexts in which all media are embedded. The example of incarcerated individuals' media work points to the role of the state in the development of media and communication technologies, as it clarifies that state-subsidized work de facto has been behind a large part of the media infrastructures that surround us. Incarcerated individuals' media work highlights the state's role in organizing, financing, and subsidizing media and media work. Third, the example of prisons highlights the need to attend to the forms of work that take place "below the line" and beyond what have traditionally been perceived as media industries. The type of work that incarcerated individuals in Swedish prisons have done has been about producing and maintaining media technologies and infrastructure required for mass communication. But as the contemporary examples of passive media work in smart prisons suggest, the correctional service can also act as an experimental workshop that contributes to the development of media and communication technologies.

4

BUILDING PRISONS: PRISON MEDIA ARCHITECTURE

The architecture of prisons carries and mediates ideas of punishment and rehabilitation. The structure of communal spaces, the individual cells, the workshops, and the outlook of the prison building itself are expressions of the imaginaries of what prisons, punishment, and rehabilitation are about. The physical appearance of the prison building obviously also directly influences the experience of incarcerated individuals and guards. It is hence no surprise that architecture features prominently in the novel *Grundbulten*. The main character repeatedly reflects on the prison experience through the architecture of Hall Prison.

> He had an epiphany, he suddenly saw before him that all the walls, all the buildings, all the culverts, all the staircases are connected in some way at Hall. They are built into each other, there is no break anywhere, Hall is one big coherent organism of concrete. (Ahl, 1974, p. 25)

> These culvert walks are of great boredom and humiliation. You walk past surveillance camera after surveillance camera. Every step is monitored. Your expressions are guarded. Admittedly, you joke about it every now and then when you go downstairs on the way to the culvert when you meet friends: "Did you put on your make-up? You know you're on TV!" But it's hollow, no one ever laughs. It's kind of an alley, it echoes where you walk between four concrete walls, one that you step on, one that you have on your head, two that press against your sides: it feels like that. At regular intervals you have to stop at green armored doors, press a button and say who you are before you are let through, even though the surveillance cameras are buzzing and watching you and even though they in the control room know very well who you are. In addition, armed guards are equipped with Walki-talkis in the door of every single workshop. Most people walk in the culvert only when they are absolutely forced to, ie. four times a day; when the prisoners are to be moved. The guards can go over ground, which they usually do. You go as fast as you can to get through as quickly as possible, the boredom and humiliation is monumental. You hate your whole environment, you hate those

> who walk next to you and therefore there is rarely any talk down there. You are on the way to a humiliation and feel humiliated and you get angry because the others know how you feel, you are about to be exploited and instead of creating resistance and solidarity you feel further humiliated by knowing that friends know you're up for this shit. (Ahl, 1974, p. 44)

The monotony of culvert systems built of concrete and the feeling of being constantly monitored became the iconic way to describe the experience of incarceration in Swedish high-security prisons of the mid-1960s. The long culvert tunnels connecting different parts of the large-scale factory prisons were supposed to ease the movement of incarcerated persons and reduce costs for staff. The idea of moving the incarcerated between buildings in an extensive tunnel system was made possible in connection with one specific technology: CCTV. The construction of modern prisons and use of the latest surveillance technology were part and parcel of reimagining Swedish corrections. The archetypical culvert systems are an outcome of one of the most comprehensive prison construction projects that Sweden experienced between 1956 and 1967, a building boom that was driven by ideas of cost reduction and efficiency through technology. In this period of 11 years, 12 new facilities—more than 1 every year—opened in Sweden following the shared idea of industrial production being at the heart of the modern penal system and modeled on the ideals of "normalization" as the dominant penal ideology.[1]

Large-scale prison construction projects are often linked to fundamental criminal justice reforms where new architectural features become both a driver for reform of the penal paradigms and expressions of new visions of the penal regime. The same was true for the period of cell prison construction (in Sweden from the 1840s and onward) that was intimately tied architecturally to new ideals of how to treat and reform incarcerated individuals. The two central periods of prison construction—the emergence of cell prisons (1846–1898) and industrial prisons (1956–1967)—are also periods in which technologies for surveillance and control were developed and incorporated in the physical prison architecture in different ways.

Furthermore, the prison architecture produces a specific communicative rhythm of prison life depending on how the incarcerated are moving between different parts for communal and recreational activities, work, and eating as well as isolation. The "microphysics of power" elaborated and developed in these institutions, for a theorist such as Michel Foucault (1979), for example, was understood as a model for how power in general operates in modern societies. The control mechanisms of the prison, the separating, categorizing, discriminating, hierarchization, and so on, were for him universal rather than specific for the institution of prisons. Therefore, the prison as a material reality is one (but not the only) answer to the question of how to create an architecture that makes possible an "internal, interconnected and detailed control" so everyone within its confines becomes visible and that could "promote the transformation of individuals, influence those it encloses, gain an understanding of their behavior, make them the object of knowledge, change them" (p. 201).

This chapter explores such entanglements of architecture, media, and the idea of reforming incarcerated individuals. Our starting point is that prison buildings are a form of media for communication. They are to a high degree solutions to problems of communication as symbolic expression and representation but also as actual movement/transportation: prisons as physical constructions are responding to questions of how to prohibit, control, manage, and (sometimes) enhance the communication of the incarcerated with staff, with each other, and with the outside world. The prison, with its walls, windows, doors, corridors, towers, screens, fences, shafts, balconies, and pipes, *is* a medium of communication. But the prison is also a medium that carries messages of crime and punishment to the wider public that crystallize the penal regimes of a given society and in that sense mediates and configures the ideas of what punishment is and what kinds of social problems the penal system should remedy.

CELL PRISON ARCHITECTURE

The first modern prison, the cell prison, is a product of the nineteenth century. The new penal regime that spread throughout the Western world during the century, with more "humane" treatment of criminals, the abandonment of corporeal punishment, and standardized practices of punishment within the nation-state, was a product of the Enlightenment and Christian revivalism and part of the general modernization of society (Forsythe, 1987; Meranze, 1996). The cell prison was designed to reform and transform incarcerated individuals and as such was a "moral technology" (Foucault, 1979) for bettering the individual grounded in a new optimism regarding the "moldability of the human" (Lundgren, 2003, p. 10). This belief was widespread in this period, a time that historian Eric Hobsbawm (1977) calls "the age of revolution" (from 1789 to 1848). "Reformatory theory" in the nineteenth century, with spiritual reformism as an important strain of thought, understood incarcerated individuals as products of their living environment (Forsythe, 1987). Hence, in order to change and reform incarcerated individuals, the environment had to be changed. The modern prison was the solution to this problem (Markus, 1993; Meranze, 1996). A novelty in these prisons was the separate cell in which incarcerated individuals would be kept for most if not all of their time in prison. Helen Johnston has pointed out that the cell was established as a "space of potential transformation in which, under the correct conditions, prisoners could reflect on their own behavior and past criminal life, repent for their crimes and look at God for salvation and a new law-abiding future" (Johnston, 2020, p. 30). In "the solitude of the cell . . . alone with God and a wounded conscience, the unhappy man is forced to exercise his powers of reflection, and thus acquires a command over his sensual impulses which will probably exert a permanent influence" (Ritchie, 1854, quoted in Johnston, 2020, p. 30).

During the nineteenth century hundreds of new prisons were constructed throughout the Western world. This was mainly an elite project, lobbied and pushed for by liberal leaders (politicians, lawyers, doctors, etc.) and liberal reformists, and not the result of popular demands for changes in penal policy. In the Swedish context, that it was an elite project is made explicit by the fact that one of the instigators behind the penal reform, a strong voice for the need to construct new prison facilities, was the crown prince and later king, Oscar I, who in 1840 authored and anonymously published the book *About Punishment and Prisons*, also known as "The Yellow Book." This publication that pleaded for a more "liberal" and "humane" treatment of criminals was part of a public debate that had been going on since the 1820s and marked the start of an intense period of prison reform and the construction of 42 new prison facilities all over the country, following a decision by the so-called liberal parliament of 1841 (Nilsson, 2013).

The reform project was, however, not as liberal as often argued. Rather, prison reform appealed to different groups for different reasons that made it viable. As the prison historian Anders Åman (1976) has pointed out, "no one had anything to lose with a prison reform. . . . [F]or the clergy it gave new opportunities for spiritual education, for administrators and conservative defenders of social order, it was a way to control a previously chaotic system, for the liberals it was an expression of the increased respect for the individual, for doctors a way to improve hygiene and control of epidemics, and for the prison staff an opportunity to increase legitimacy and status" (Åman, 1976, p. 625).

When it was decided that large investments in Swedish prison infrastructure would be made, the fortification commander at the garrison in Kristianstad in southern Sweden, Carl Fredrik Hjelm, was appointed as the main architect of the projects. He later became head of the construction offices within the Swedish Prison and Probation Service (at the time known unofficially as the board of the country's prisons and work establishments) and was the mastermind behind many of the new facilities built in Sweden in the nineteenth century. In 1842 and 1843 he traveled around Europe to inspect new prison facilities, mainly in England, France, and Belgium, to find inspiration for the projects at home. Upon his return to Sweden, he published his impressions in a long article in the newspaper *Aftonbladet*, which carefully accounted for the technical refinements and architectural principles of foremost the British prisons he had visited (Hjelm, 1848). His work led to the design of two model facilities: one larger T-shaped building with 66 to 102 cells and a smaller rectangular building in three floors containing up to 54 cells (figure 4.1). Hjelm himself drew all the prisons constructed in Sweden until 1855, when he left his assignment at the the board of the country's prisons and work establishments.

The first modern cell prison is usually considered to be the San Michele prison in Italy, constructed by Carlo Fontana and commissioned by Pope Clemens XI (1704), and the Gent prison drawn by the architect Malfaison (1775). The principles used to construct these facilities were described by the prison reformer John Howard (1777) in his book *The State of the Prisons in England and Wales, with Preliminary Observations and an Account of Some Foreign Prisons*. Howard and Jeremy Bentham, prison reformers of the time, tried to construct a model prison built on new and more humanist principles of reform. The importance that Bentham put on the buildings and the possibilities to solve social problems through physical constructions is neatly summarized in the first sentence of his famous book *Panopticon* in 1791: "Morals reformed—health preserved—industry invigorated instruction diffused—public burthens lightened—Economy seated, as it were, upon a rock—the gordian knot of the Poor-Laws are not cut, but untied—all by a simple idea in Architecture!" (Bentham, 1791/1995, p. 30).

Fängelse

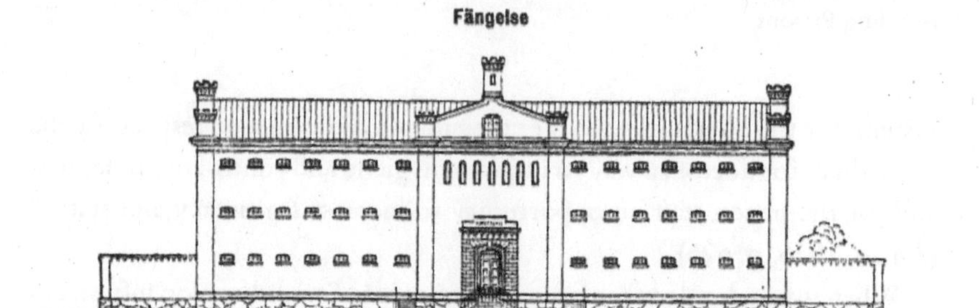

Fig. 1. Fasad. (Det framskjutande midtpartiet inrymmer administrationslokaler.)

Fig. 2. Gafvel och det framskjutande midtpartiet.

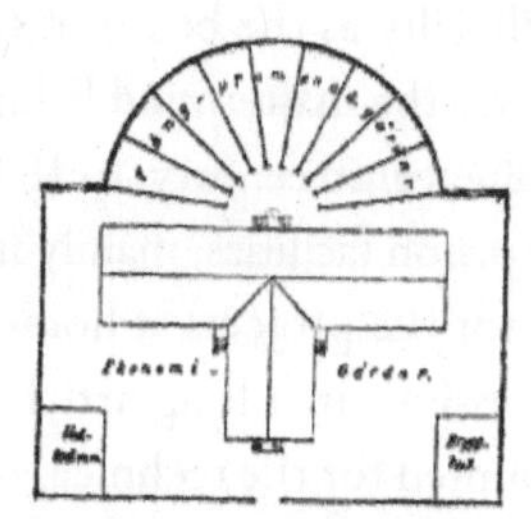

Fig. 3 Situationsplan af fängelset.

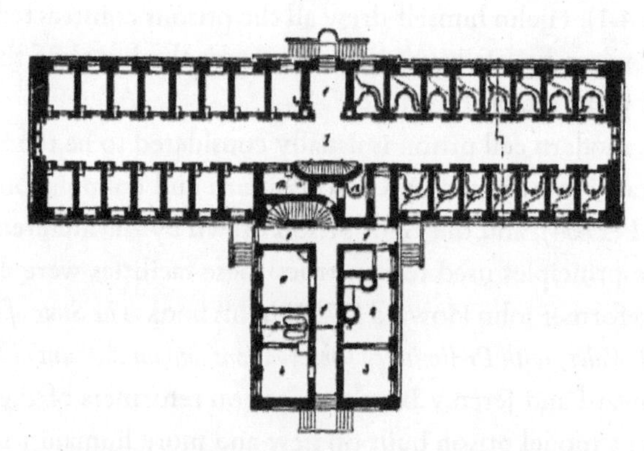

Fig. 4. Plan af mellanvåningen, visande cellernas läge m. m. *1.* Förstugor och korridorer. — *2.* Kök. — *3.* Kokerskans rum. — *4.* Mottagningsrum. — *5.* Badrum. — *6.* Vaktrum. — *12.* Mörka celler. — (*S*, *T*, *U* och *W* beteckna detaljer af värmelednings- och luftväxlingssystemet.)

Svenskt länscellfängelse (resp. kronohäkte).

Figure 4.1
One of the two model type prisons designed by Carl Fredrik Hjelm (1793–1858) and later constructed, with architectural adaptations, in all Swedish major cities between 1846 and 1898. Drawings from Nordisk Familjebok, Public Domain.

It was not until the nineteenth century, however, that the American architect John Haviland constructed what is considered the first true modern penitentiary. Accordingly, the Eastern State Penitentiary in Philadelphia was built in a way to facilitate solitude and silence and had, for example, no communal spaces for work activities. This ideal facility became the model for what is known as the Philadelphia system. The Philadelphia system, envisioned and developed by the Quaker movement in the US, eventually became the main reference point for the first modern facilities in Sweden and around the world.

The dominant form of the nineteenth-century modern prison was the radial design that divided incarcerated individuals into different groups with a central point in the middle where control and surveillance were concentrated, an idea most vividly developed by Jeremy Bentham (1791/1995) in his work on the panopticon. The watchtower, which was ideally not only the central control room but also the living area of the warden and his family, was the core of the prison that functioned as a "depersonalized form of regulation built into the structure and design of the institution" (Matthews, 2009, p. 32). The control from the watchtower overlooking the whole facility and offering insight into every cell was independent from any particular controller, never seen but constantly feared. In that way the controlled, the incarcerated individuals, became the bearers of internalized, assumed constant control that transcended the traditional power relations within the prison that were based on the opposition of coercion and consent (Matthews, 2009). Foucault famously argued that the "architectural apparatus should be a machine for creating and sustaining a power relation independent of the person who exercises it; . . . the inmates should be caught up in a power situation of which they are themselves the bearers" (Foucault, 1979, p. 201).

The cell prisons that were constructed around the world at this time share many similarities both in style and the technical solutions used in order to realize the new penal ideals. The discussions on the technical issues were lively, and many complicated and refined systems were imagined for how to turn these facilities into effective machines for reform (Lundgren, 2003, p. 68). The building style was monumental and weighty, and as Åman (1976) has argued the use of historical shapes in prison architecture could be interpreted as an illustration of the social function of the prison, for example, leading to associations with castles and fortresses that are symbols and expressions of power. As Hancock and Jewkes (2011) argue, nineteenth-century prison buildings grew more expansive and grandiose; "they communicated a clear message about the perils of crime and the uncompromising nature of the State's retribution" (p. 616). Similarly, the architect for the city prison in Stockholm made the following comment on the style of the building: "For the building's exterior I have chosen a strong and austere, though not gloomy or discouraging character, approaching the Florentine medieval style" (quoted in Åman, 1976, p. 110).

As Åman also suggests, the neoclassicist and later neogothic building styles were fashionable and typical for the time, seen in much contemporary architecture, and it might be wise to "be careful with conclusions about their symbolic content" (Åman, 1976, p. 119). The Swedish politician Carl Lindhagen in a critical remark even referred to the cell prison as one of the three architectural "superpowers" of the time, together with the factory and the military barrack (Lindhagen, 1929, quoted in Knauff, 2012, p. 89). Observations of the similarities between these buildings of modernity is common and perhaps most famously phrased by Foucault (1979), as he rhetorically asks "is it surprising that prisons resemble factories, schools, barracks, hospitals, which all resemble prisons?" (p. 228). The principles laid out by Jeremy Bentham (1791/1995), for example, are explicitly not only about prison architecture but also represent a "new principle of construction [for] any sort of establishment, in which persons of any description are to be kept under inspection" (p. 29). As shown by Crawford (2021), Bentham was inspired by the factories owned and managed by his brother Samuel Bentham (working under Prince Potemkin in Russia) when developing his influential idea of the panopticon. This underlines the generality of the ideas of control and communication—manifested in the panopticon—for modernity and modern organizations and bureaucracies. The idea of solving social problems with architecture and with the control of communication in and through physical space was (and is) a modern idea applicable to numerous operations.

When Carl Fredrik Hjelm was appointed architect for the new cell prisons in Sweden, he was given instructions to specifically develop the technical solutions in the prisons, and they were considered to be advanced, even luxurious at the time. Not only were their exteriors described as "palace-like," but the solutions to problems of heating, isolation, and lighting, which introduced many innovations in construction engineering, also gave the new cell prisons an air of exclusivity.[2] When the cell prison in Linköping, a town in southern Sweden, opened in 1856, for example, the local press reported that with such prisons "it will certainly become more advantageous to be a thief in Sweden, than to be an honest man" (quoted in Blomqvist & Waldetoft, 1997, p. 26). Even though many of the more advanced technical solutions and innovations in construction engineering never left the drawing board, the architecture was indeed innovative and represented new ideas of reforming incarcerated individuals. The physical and mental dimensions of the incarcerated were not separated in the discussions of the prison at this time. Rather, "body and soul were related as corresponding vessels" (Lundgren, 2003, p. 41), and the architecture conjoined these dimensions and used the physical structure of the facility to cater to habits, rhythms, and sensory impressions.

In the nineteenth century architecture, as argued by Michael Meranze (1996), "had moved to the center stage in organizing prison life" (p. 250), and one of the key issues for prison architecture was the problem of communication. The prison environment was supposed to prohibit certain forms of communication and interaction in order not to become a "school of vise" (p. 225) and to enhance possibilities for surveillance and security. As Bentham had previously argued on the panopticon, the incarcerated individual should be "secluded from all communication with each other" (Bentham, 1791/1995, p. 34). John Haviland in this respect referred to the "evil of conversing": the problem of unconstrained communication between the incarcerated individuals (quoted in Meranze, 1996, p. 249). On the other hand, the prison architecture should enhance other "proper" forms of communication and mediation. Michel Foucault recognizes and emphasizes these communicational aspects of the prison structure. He argues that "each individual, in his place, is securely confined to a cell from which he is seen from the front by the supervisor; but the side walls prevent him from coming into contact with his companions. He is seen, but he does not see; he is the object of information, never a subject in communication" (Foucault, 1979, p. 200).

One aspect of prisons as communication infrastructures is the production of visibility and invisibility. The modern cell prison of the nineteenth century introduces separation and seclusion, needed for monitoring and control for moral reform. Within the prison, incarcerated persons were separated from each other while they were removed from the public eye and from the rest of society. From the opening of the first cell prison in the Swedish town of Falun in 1848, newspaper reports tell that crowds had been gathering in the streets to witness how the 45 incarcerated individuals were moved from the old facilities to the new prison building. It was a short walk between the two buildings, but the procedure took several hours and became somewhat of a public spectacle. This, however, was also the last time that incarcerated individuals were seen, in this form of public performance, in the town. As the prison historian Jörgen Gustavsson (1989) remarks, this event marks a breaking point, and from then on "there was only a handful people that ever came into contact with the incarcerated individuals" (p. 64). Those people with access to this specific prison, as in most other cases, were the warden, the priest, the doctor, and a small group of prison guards. As the incarcerated individuals became increasingly invisible, the institution itself, the prison, was more visible than ever before. The new facilities were erected in the city centers and were large buildings, often interpreted in contemporary commentary as "castles," as in a press commentary on the opening of the cell prison in Linköping in 1846: "The building and the construction work are beautiful and, the windows excluded, extremely palace-like. If a traveler, unfamiliar with the city, were to seek his way to the castle by himself, he would certainly stop at this prison and ask if the Count was at home" (*Östgöta Correspondenten*, June 6, 1846, quote from Blomqvist & Waldetoft 1997, p. 26).

From now on the incarcerated persons themselves, previously visible in the city as part of their publicly performed corporeal punishment, were no longer the main messenger of the consequences of crime: the institution as such, with its high visibility, became the symbol of the relation between society and its deviants. The seclusion and separation from the public were also carefully considered in the architectural details of the construction. For example, the placement of windows and the ventilation of the facility puzzled the architects and engineers, since the necessity of opening the windows, at least during summertime, made it possible for the incarcerated to communicate with the outside world and also made them visible for people passing by. Fences and palings became the solution to these problems and also at times prohibition of ventilation through opening the windows.

The new facilities not only produced a specific and ambiguous invisibility of the incarcerated individuals in society; the general idea was also that they should be invisible from each other within the prison. The belief was in part that solitude was a precondition for moral reform (through introspection), but it also mirrored the idea of crime as a (social) contagion, which made "infection control" an important part of the idea and architecture of the modern cell prison. Any contact between incarcerated individuals could lead to spreading the social virus of criminality and therefore architectural efforts as well as efforts of interior design, and different technical solutions were taken to make the incarcerated individuals invisible to each other. For example, the cells were placed asymmetrically in the corridors so that an incarcerated person from one cell could not see into the opposite cell in the unlikely event that both cell doors were open at the same time.

The most famous of these technical arrangements was the cell cabinets that were used to keep incarcerated individuals out of each other's sight during communal activities including teaching and religious services. Booths were designed in which incarcerated individuals would sit or stand next to each other but unable to speak to or see each other. In this way they could participate in collective gatherings without any—or at least very restricted—possibilities for communicating. Another infamous construction was the separation walls at the prison yards, which made it possible for incarcerated individuals to get the one hour of outdoor stay per day stipulated by law without seeing or coming into contact with each other. In cases when incarcerated individuals had to move or be moved through the communal areas of the prison, another device was used to facilitate their invisibility, namely prison masks. This is not an architectural feature or a part of the interior design, but it was a technology used in order to restrict the vision of incarcerated individuals and allow for them to be "invisible objects" within the prison.

A more direct effect of the prison architecture with consequences for the possibility to see was the lighting in the prisons. As it was decided that the new prison facilities should have central heating and not fireplaces, the main source of lighting disappeared from the cells. In the 1840s and 1850s, paraffin lamps did not exist. They were invented in 1854 and began to be used more broadly in Sweden first in the late 1860s. The prisons had only small windows, and during wintertime daylight is a scarce resource in the Nordic countries. This meant that prisons were generally dark, and many accounts from the time speak of the poor lighting conditions. The many hours of darkness in the cells were considered to be an unfairly harsh treatment for incarcerated individuals but also detrimental to the health of the employees working in the prison facilities (Blomqvist & Waldetoft, 1997, pp. 49–50).

While the prison architecture, the interior design, and various technologies in the prison in this way produced invisibility, the prison also was a site that produced forms of visibility. Incarcerated individuals were to be made visible for the guards, for example, as objects of surveillance and control. The idea was also to make incarcerated individuals visible to themselves through solitude, silence, and isolation and thereby heighten and intensify introspection so as to form the incarcerated as individuals with amoral responsibility, self-knowledge, and self-control. An important aspect of the architecture and interior design of the cell prison was that it was supposed to have, as the document from the parliament prescribes, a "stripped down decor so as not to be distracting" (Gustavsson, 1989, p. 62). What the policy makers were afraid of was the distraction from the attention to oneself, the self-communication that incarcerated individuals should devote themselves to in order to promote a "curing of the soul."

Visibility was again achieved in part through architecture. This was a key idea in the notion of the panopticon but also in the types of prisons that were actually realized. The centrally placed watchtowers and the galleries allowed for surveillance of incarcerated individuals and constant visibility from guards. The architecture and interior design also produced visibility in other ways, such as slates outside each cell door with information (for the guards) about the incarcerated individuals and the reversed door-eyes in each cell door that allowed for the guards to see through cell doors without the knowledge of incarcerated individuals in their cells. Vision and sound were regulated and managed through the architectural features of the cell prison. Silence was mandated for the incarcerated individuals as well as for the guards. The very first paragraph of the handbook for Swedish prison directors from the time states that "within a cell prison, silence and stillness must be observed" (Mentzer, 1878, p. 19). In the architectural design of the prison, the issue of silence was a problem that needed solutions. The walls between the cells, for example, needed to be thick enough to be soundproof; on the other hand, some sound needed to be able to penetrate the construction, since an important part of the surveillance of the incarcerated individuals was to hear them in case they broke the order of silence and consequently punish them.

While silence was the most common mode, enhanced by the architecture, at times the silence was broken, and incarcerated individuals were supposed to listen. For example, the incarcerated were allowed to have private conversations with a priest, and each Sunday a religious service was held at the prison. The prison congregation was taking part in the service from their cells; the cell doors were opened about two inches and locked in this position by a hook. The priest held his service from a pulpit designed especially for this purpose that was placed between the balconies, in the middle of the gallery, so he could be heard by everyone in the building. The participation was mandatory, but as the priests often complained, the fact that the incarcerated individuals were not visible during the service and not allowed to participate in the singing of the hymns made it difficult to supervise whether they were taking part of the service or not. This protobroadcasting of religious services to incarcerated individuals was replaced in the twentieth century with actual broadcasting, as the priest was replaced with the radio broadcast of the Sunday mass in many of the Swedish prisons.

Nevertheless, the building itself also became a medium of "alternative" or "oppositional" forms of communication between incarcerated individuals. As preserved artifacts (often seized by the guards) and archival sources reveal, incarcerated individuals showed a great deal of creativity and were willing to take considerable risks in order to facilitate (the strictly forbidden) communication between them. Notes, letters, poems, and drawings made with contraband were passed around in inventive ways, and even though all contact between the incarcerated was prohibited the records show that such contacts were common. There are even documented cases of love affairs blooming in the ordered isolation, leading to marriages after release (at this time men and women were generally not separated in different facilities).

In this respect the architecture itself, the physical structure of the building, became an important medium of communication between the incarcerated individuals. The prison management could of course try to prohibit and punish this behavior, but in contrast to contraband such as writing equipment, they could not seize tools for communication that were part of the architecture. Writing on or carving messages into the walls was of course one possible mode of expression and communication, even though the message could not be received by anyone except the next tenant in the cell.

Kate Herrity (2020) has pointed out that sound is an important element of communication and surveillance within the modern prison. Guards are listening to the incarcerated individuals, and from the overall soundscapes in the facility guards can anticipate eruptions of violence. The sounds they themselves make, such as their shoes on the floor or the sound of their keys as they move around the prison, are interpreted by the incarcerated as impositions of symbolic power (Herrity, 2020, p. 246). The incarcerated themselves also use the constructions of the prisons (doors, pipes, etc.) to make sound through which they can express themselves and communicate. Banging on doors and knocking on walls and pipes are key elements of these forms of communication. Banging, Herrity (2020) argues, "represented an act of insistent communication in opposition to the constraints of the physical environment and in that sense constituted an act of resistance with a variety of discernible messages: a system of meaning" (p. 242). Since almost all modern prisons in the nineteenth century had systems of central heating, the pipes became effective carriers of sound between the cells. At least three different variants of the "knocking alphabet" (Blomqvist & Waldetoft, 1997, p. 115) have been described and preserved, but most probably more variants existed, and all of them used a rather high number of knockings for each word, which means that both patience and attention were required when communicating in this way. The fact that the guards passed each cell every third minute every hour of the day and night made the use of knocking even more difficult, but there exist numerous accounts in the prison archives of how incarcerated individuals used this alphabet to coordinate action, plan smuggling of contraband, and gain "forbidden information from their comrades" (Blomqvist & Waldetoft, 1997, p. 116).

Communication and mediated surveillance were key elements of the architecture of the modern, nineteenth-century prison. Architecture was understood as the means for solving the social problem of criminality, and that was done through controlling and steering communication and creating an informational architecture that makes incarcerated individuals visible and invisible: hindered as well as facilitated communication. The ideals and imaginaries underpinning the modern prison became, however, increasingly questioned, discredited, and abandoned during the late nineteenth and early twentieth centuries. The architectural manifestations of those ideas nevertheless remained intact and in use well into the twentieth century. Helen Johnston (2020) has pointed to the path dependency that the construction of the modern cell prisons facilitated within penal policy, and she quotes one chairman in the British prison commission in the mid-twentieth century who stated that "nothing less than dynamite could rid us of these grimy and forbidding fortresses" (Johnston, 2020, p. 41). After the war, however, there was a reorientation in penal policy and a new wave of prison construction, and a new era for prison architecture came into being as old facilities were increasingly abandoned and new prisons emerged.

INDUSTRIAL PRISON ARCHITECTURE

The modern cell prison was the dominant architectural form until a larger set of reforms of the penal system in Sweden took shape during the 1940s. The reforms led in subsequent decades to the abandonment of the cell prison and the construction of a new generation of prisons. The emergence of new prisons in Sweden was again part of a broader international trend. The rise in prison populations in addition to a new penal ideology, new therapeutic ideals, and new architectural values and styles as well as new building techniques, manifested themselves in these new facilities in Sweden and internationally. Prisons have always been "structures for reshaping human character" (Luckhurst, 2019, p. 159), and the new types of industrial prisons from the postwar period onward are no exception. A report by the Royal Building Committee summarizes the approach to corrections that was supposed to be manifested in the new facilities:

> In many respects, the proposals put forward by this committee were characterized by a completely new approach to corrections. Thus, it proposed facilities with considerably increased medical resources, residential wards for smaller groups of inmates, increased flight security, industrial work, etc. Not least, the new work operations changed the planning and appearance of the prisons compared to the old solitary confinement facilities. Large areas of land were required for workshops, warehouses and traffic equipment, which ruled out the idea of building on top of the older prisons, usually squeezed into narrow city centers. There had to be entirely new facilities. With the advanced workshops and access to tools and working materials also came difficulties in making the facilities secure against flight attempts and escapes. It is no exaggeration to say that the need for security weighed heavily in the planning and design of systems and details. (KBS Rapport, 1969)

The new facilities were also considered to be more humane and to represent the new ideals of "normalization"; that is, prison life as much as possible should resemble ordinary life outside of the prison. Taken to its most extreme, as in the plans for the never-realized prison in Uppsala, the incarcerated should live in what resembled small townhouses, constructed in such a way that the prison wall would not be visible for them most of the time. Incarcerated individuals were supposed to spend much of their time together with each other, far different from the corridors, cells, and solitude of the nineteenth-century prison institution.

In 1956, a parliamentary committee was formed to "handle new construction activities in the Prison and Probation Service" (SOU, 1959:6, p. 7). Sweden was not the only country that developed and built new prisons at this time; on the contrary, this movement was part of an international trend, and the parliamentary committee followed carefully the development in other countries especially the United Kingdom, France, and the US. Hancock and Jewkes (2011) comment on the international prison development in this period as "resonat[ing] with an important aspect of much organizational design, promoting a high-modernist rejection of the aesthetics ornamentation or decoration," and instead turned to "what was considered to be humanely functional, styles of high, progressive modernism." (p. 617), which must be said to hold true also for the Swedish development in this era. The head architect in the committee was Tor Bunner, who saw architecture as an important source for social change and transformation:

> It is indisputable that we today are in the midst of an innovation of the methods and principles for treatment of the clientele in the Prison and Probation Service that are as radical as those who created the prison system in the 19th century. . . . The new ideas and programs for treatment have not yet found their physical form and we therefore have to expect that the drawings for each new prison, which is constructed today, quickly will be obsolete. It is reasonable to view the current period as an experimental phase during which a large number of suggestions and ideas will be tested and evaluated. It is only in this way that the new prisons can make a difference for the social rehabilitation of criminals. The creation of the right physical environment that contributes to this rehabilitation must become the tribute of architecture in the fight against crime. (Bunner, 1967, p. 642)

Until World War II the prisons in Sweden had a steady population of about 2,000 incarcerated individuals per year, but this changed after the war, especially during the 1950s. This is another background to the urgency with which this matter was handled by the authorities: the prison population doubled during the 1950s to around 4,000 incarcerated individuals. And the projections at the time were that the prison population would increase even more and double once again during the following decade and reach as many as 10,000 incarcerated individuals by the 1980s (Åman 1976, p. 403).

The requirements in the new policy from 1945, forming one of the cornerstones in the Swedish idea of normalization that included daily routines that resembled life outside the prison as much as possible: eight hours of work (or studies), eight hours of leisure, and eight hours of rest. The facilities needed to be designed so as to make this ideal a reality. Space was needed for large industrial facilities (the factories) as well as for leisure (sporting fields, libraries, auditoriums, etc.). Another important dimension in the new penal ideology was the "principle of the small group" (Fångvårdens byggnadskommitté, 1965). The rehabilitation of incarcerated individuals was thought to be enhanced if the incarcerated person's life could be organized within a smaller community of other incarcerated individuals and staff. Within a smaller group, relationships based on trust could be fostered over time, which would be beneficial for the "treatment" and "care" of the incarcerated people. If the previous era of prison architecture had privileged silence and solitude, the industrial prisons tried to enable communication and community. The small-group principle was also beneficial for security reasons, making it easier to monitor incarcerated individuals as well as to "prevent uncontrollable contact between the incarcerated individuals at the closed institutions and the outside world" (Fångvårdens byggnadskommitté, 1965, p. 14). The solution to these issues was the construction of larger prison areas placed far outside the city centers, where there was enough space and cheap land available. As argued by Anders Åman (1976), the "architectonical prototype for the new prisons of the 1950s and 60s was the camp, the war prisoner camp or even the concentration camp: a larger number of small buildings, distributed over a rather large area surrounded by a fence or a wall" (Åman 1976, p. 398).

In the Swedish case, the areas were in general surrounded by high inward-leaning walls of concrete that were initially five meters high but later heightened to seven meters. The increase of wall height was an initiative from the director general of the Swedish Prison and Probation Service at the time, Torsten Eriksson. He supported this decision with the possibilities for free movement within the institution. At a meeting with the Building Committee in 1965, he argued that "a seven-meter wall . . . seems to be the best solution. If such a wall exists, also a clientele that is prone to escape can be allowed relatively free movement inside the walls, given that the facility is not too large" (quoted in Gahrton, 1971, p. 81). Inside these walls, several smaller one-story houses were scattered. There were barracks for housing, in general brick houses, sized to be suitable for a "small group" (figure 4.2). There were also larger concrete buildings, the industrial facilities. These were highly modern spaces of production. For example, when Kumla Prison was opened in 1965, it hosted the most advanced and state-of-the-art industrial laundry in Europe. Since the new penal policy from 1945 had explicitly mandated that incarcerated individuals should have possibilities for sport and exercise, sporting fields were a part of the new facilities as well as "ping-pong tables at every ward" (Fångvårdens byggnadskommitté, Slutrapport 1972, p. 6).

The Prison and Probation Service's Building Committee launched a report in 1959 titled "The Optimal Size of a Prison" in which these new principles were laid out. The report concluded that 450 incarcerated individuals were an optimal number for a larger facility and 150 for a smaller one. This would also make the prison economically sustainable while upholding the principle of the small group and effective industrial production. Furthermore, since all of the new facilities were planned with modern factories and industrial facilities within the walled area, the committee elaborated the idea that the profits from industrial production should be used to pay market-based

Figure 4.2
Housing area at Kumla Prison 1965. Uppsala-Bild/Upplandsmuseet. License: Attribution-Non Commercial—No Derivatives (CC BY-NC-ND).

salaries to incarcerated individuals so that "with their salaries to some extent [they] can compensate the state for the costs of their imprisonment" (Fångvårdens byggnadskommitté, 1965, p. 35). All in all, larger facilities of this kind with advanced industrial production, the committee concluded, are far more economically efficient than smaller ones. The Building Committee estimated that as long as the incarcerated individuals are "differentiated" in small groups and in general only meet and socialize during work hours, the negative effects of larger facilities, namely the fact that prison is a "school of vice" where younger incarcerated individuals meet and learn from their older fellows, can be avoided.

The report on the "optimal size" of a prison was met with heavy critique. The critics especially pointed to the fact that the committee mainly had taken the economic aspects into consideration and neglected the therapeutic or social consequences of large-scale high-security facilities of the kinds suggested in the report. Somewhat surprisingly, then, during the 1960s and 1970s the committee made no attempts to study or get a better understanding of the relationship between the architecture and outline of the facilities and the social and therapeutic dimension. In its final report, published in 1972 when the Building Committee was disbanded, the committee briefly discuss the "further need" of such studies and urged future policy makers to take these aspects into consideration (e.g., conduct a survey of these matters). The report states that "such a study would clarify to what extent construction engineering can contribute to giving the atmosphere of the institutions a more personal and therefore care-oriented touch" (Fångvårdens byggnadskommitté, 1972, p. 49). Even though the critique of the report was widespread, the general ideas of the committee laid the groundwork for the development of new facilities during the following decades. The report on the optimal size of a prison indeed was one of the initial "foundation bolts" for the Swedish prison system developing in the welfare state of the postwar years.

THE "BUNKER PRISON" AND ITS CRITICS

Prisons, along with other large-scale institutions of the nineteenth century such as reformatories, hospitals, factories, workhouses, and asylums, according to the historian Roger Luckhurst, distributed space in ways that were fundamentally "corridoric" (Luckhurst, 2019, p. 159), namely built around central corridors that connect the different rooms of the building. Corridors as such are an important mediating space between different rooms. The term "corridor" is derived from French and Italian *corridore* and *corridoio*, respectively, for "running place," a passage you are supposed to move through quickly. The term is also related to correspondence and hence has a clear communicational grounding. The first corridors were built in the fourteenth century in the Spanish and Italian context before the word "corridor" in Latin referred not to a space but instead to a courier, a messenger who could run fast to, for example, deliver secret messages during wartime to the front line or an independent negotiator who arranged trades between families. By the seventeenth century the word "courier" was predominantly used to describe such services, and the word "corridor" was mainly used to describe military forms of fortification that allowed for quick communication between different troops (Jarzombek, 2010). *Corridorio* could also be used to describe secret passages in and out of a castle or palace. The corridor was hence connected with the notion of speed and quick and efficient communication, but it was also connected with the image of being dark and lonely, even haunted, which led to a resistance against corridors in public architecture of early nineteenth-century England. The early modern form of corridors emerged in European architecture in the 1770s in connection with military buildings for the cavalry regiments that became the first purely corridor structures and formed the prototypes for modern corridor buildings. Corridors quickly became standard elements of state buildings including universities, town halls, and other governmental buildings in the US and Europe and constituted "an instrument of surveillance, channeling and dividing people into its spatial regimes" (Luckhurst, 2019, p. 171).

Corridors in general but especially in the prison context do not necessarily form an opening between the inside and the outside. Rather, they constitute a passage within a closed architectural system. However, they also function as architectural media similarly to doors and gates as they mediate between different rooms and activities (Siegert, 2015). In the prison context, corridors connect the communal and living areas with the workshop, the library, and the exercise yard. But to a large extent, corridors are nonplaces that are experienced as fundamentally alienating while also from a more practical standpoint being criticized for their bad ventilation and tendency to contribute the spread of infections (Luckhurst, 2019). All such familiar critique toward corridors was also manifest in the discussions of what was critically labeled as the new "bunker prison" that focused on the alienating and antihuman aspects of the industrial prison architecture and to a large extent was what spurred the larger and intense societal debate on prisons and penal care that culminated with strikes and other forms of political action among incarcerated individuals in the 1970s (Adamson, 2004).

In a sharp attack on the new prisons, the prison psychologist at Kumla Prison authored a report published in Sweden's largest daily, *Dagens nyheter*, in 1966. The report concluded that incarcerated individuals were tormented by claustrophobia and severe apathic conditions, spending most of their time inside and underground (in the culvert system) behind seven-meter walls, rarely seeing daylight and lacking personal contact with the staff (quoted in Paulsen, 1971). One incarcerated individual quoted by the daily *Aftonbladet* argued that the number of suicides will rise tremendously in such a clinical and alienating environment: "Everybody who is placed here [at Kumla] is getting more or less mentally sick. The prison is like a closed tin can—the incarcerated individuals are like pickled herring" (Appelgren, 1967). The same article illustrates the experience of dehumanization at Kumla Prison with a picture of the extensive culvert system (figure 4.3).

The culvert system and the security wall of seven meters at Kumla Prison were considered especially alienating by critics and the incarcerated. One

Figure 4.3
Tunnel system of Kumla Prison, 1956. Uppsala-Bild/Upplandsmuseet. License: Attribution-Non Commercial—No Derivatives (CC BY-NC-ND).

incarcerated person is cited by the Swedish daily *Göteborgsposten* in 1966, almost two years after the opening of Kumla in February 1965:

> The wall probably looks more sinister from the outside than from the inside, but here everything is like a wall. Despite the large area of the institution (large, beautiful spaces between the houses and the wall), we spend far too little time in the fresh air. You walk from your room (they don't say cell in Kumla!) to the workshop underground, in a long desolate culvert. It would be nicer if you could take that walk above ground and out in the air. You're locked in all the time behind one big wall, it's all one wall. (Rollo, 1966)

In an op-ed in one of Sweden's largest dailies, the conservative *Svenska Dagbladet*, from 1966 (following up on the report from the psychologist at Kumla Prison mentioned above), the concept of the bunker prison was formulated, and the critique of the bunker was fierce: "It is a technical monster in steel and concrete, mastered and controlled with precision from maneuver boards and control rooms [and is an example of] ice-cold functionalism" (*Svenska Dagbladet*, February 26, 1966, quoted in Paulsen, 1971).

The idea of the new prisons as bunkers took off and became a symbol for the heterogenous movements of prison critique and reform that was forming toward the end of the 1960s. A major starting point for this movement was the so-called thief-parliament that conjoined in the Swedish town of Strömsund in 1966 where former incarcerated individuals, politicians, professionals, and employees of the Swedish Prison and Probation Service gathered for a conference that ended in the adoption of a program of principles for prison reform. The architecture of the new prisons was one of the major focal points in the critical debate on penal care. At the core of the critique was the discrepancy between the therapeutic ideals of resocialization and rehabilitation of the incarcerated (which was the official doctrine) and the possibilities for realizing these principals within the physical environment of the bunker prison (Åman, 1976). In one of the speeches at the conference, the journalist and prison activist Jörgen Eriksson stated that

> the Kumla Prison and the planned facility in Österåker is the "new deal" in Swedish penal policy. . . . With its rigorous confinement and shielding from the outside world and with its drill-like forced work, the Kumla Prison is a big leap backwards in humanitarian terms. The huge scope of the institution also complicates each attempt at individual care and treatment. Behind this new deal is a primitive, moral conception of man. (Eriksson, 1966)

Liberal politicians, professionals (including doctors, psychologists, social workers, and employees within the Swedish Prison and Probation Service), prison activists, and incarcerated individuals themselves joined in this multifaceted "prison struggle," demanding reforms of the Swedish prisons that was lively and ongoing for most of the late 1960s and the 1970s and channeled through the organization KRUM that worked for the humanization of the Swedish prison system. This critical debate on the new modern prison facilities led to several public investigations by both internal and independent committees. One such investigation was a large-scale survey among incarcerated and staff at Kumla conducted in 1969 by Göran Ekblad and the architect Sven Thiberg at the Public Institute for Construction Research. The survey and report were motivated by the status of Kumla as an experimental facility that was supposed to serve as a model for numerous other prisons across Sweden while also being the largest facility. The research project's main goal was to evaluate whether the reformist ideas behind the new model prison had been realized, namely if the goal of rehabilitation and normalization was reached. Their research focused primarily on the use and evaluation of the different buildings at the facility including the communal, living, and working areas. The fieldwork was impacted by a hunger strike that began just a week before the questionnaires were distributed at Kumla. The open-ended questions, however, confirmed parts of the public debate on the alienating character of the facility, as this answer to one of the questions illustrates, "The freedom within the prison is too restricted. Since all movement takes place within a closed unit, all this surveillance seems like something of a catwalk. Each small sign of surveillance consciously or unconsciously reminds you of the situation you are in. And it's sometimes very nice to be able to forget about it and try to push away thoughts of freedom" (Ekblad, 1967).

The critique of the alienating and antihuman architecture of the industrial prison also resonates in prison literature from the time. The following poem, written by an anonymous incarcerated person in the 1970s and published in a prison paper, reads as an indictment toward the Prison and Probation Service's Building Committee:

A desperate man,
Björn Holmstrand,
longing for freedom
blew himself out through the cell-window.

His legs broke
on the concrete
while the dawn slowly rose
over the jail
and its three hundred new detention cells.

The year is nineteenseventyseven!
The Prison and Probation Service's
Building Committee
is still
on the loose!
-amen![3]

A reaction to the critique of the antihuman environment in the modern industrial prisons was to try to modify the appearance of the gray concrete walls and culverts. One such initiative, at the Österåker prison, was an art project that covered the culvert walls with paintings by incarcerated persons. The project and the content of the art on the culvert walls were discussed in the issue of the staff magazine for the Swedish Prison and Probation Service in 1976:

> What is certain, however, is that the art of many of these walls must be described as symbolic: a wall that collapses, a ladder against a window and birds against a blue sky. . . . These are symbols of freedom painted by incarcerated individuals at the Österåker facility in collaboration with art college students. (*Kriminalvården: Tidning för Kriminalvårdsverkets personal*, March 1976)

And one of the incarcerated individuals' comments on the initiative in a positive spirit in the same article in the staff magazine: "We really like this, simply because we experience this painting job as much more meaningful than all production-oriented tasks that we normally have to do. Here we actually do something constructive for our own environment" (*Kriminalvården: Tidning för Kriminalvårdsverkets personal*, March 1976).

The high walls, the culvert, and the concrete but also—and more importantly—the technical surveillance (CCTV) was at the core of the critique toward the antihuman environment of the industrial prison. To some extent this critique must have been surprising to the director general of the Swedish Prison and Probation Service, Torsten Eriksson. Since the 1930s he had been a strong voice for prison reform and for more humane and therapeutic penal care. In the 1940s and 1950s, he was one of the pioneers in shifting the Swedish Prison and Probation Services toward a more liberal direction. If anything, he was accustomed to critique from conservative voices that the modern prison and probation service was too soft and that prisons were too comfortable and luxurious and not focused enough on deterrence. In a special issue of the industry magazine *Arkitektur* from 1967, devoted to the new industrial prisons (and in which Eriksson himself contributed an article), it is also argued, for example, that the modern and large-scale prison facilities were part of a shift in penal policy in alignment with the "democratic . . . humanism and sense for human dignity" and that architects in Sweden had been aware of their long-term social and political responsibilities, since the new facilities aim to "quickly educate individuals to adapt to a normal social life" (Olsson, 1967, p. 609). Eriksson tried to defend the Swedish Prison and Probation Service as well as the new prison architecture publicly in public lectures, newspaper articles, and a book in 1967 titled *Penal Care: Ideas and Experiments*. Nevertheless, the heavy critique toward the large, centralized high-security industrial prisons led to a reorientation in prison architecture and development. In a fierce attack on the new prisons and on the Building Committee published in the magazine *Pockettidningen R*, the prison reformer and liberal politician Per Gahrton (1971) called it "a study in organized irrationality," and the new attorney general, the Social Democrat Lennart Geijer, in 1970 publicly promised that no further development of large-scale prisons would be implemented. Several of the planned facilities were stopped and were never constructed, some of the others (such as in Österåker) were downsized, and the building committee, of which Eriksson was the chair, was discontinued in 1972.

COMMUNITIES AND CULVERTS: THE COMMUNICATION PROBLEM IN THE INDUSTRIAL PRISON

Even if the building committee was discontinued and prison construction in the 1980s took a new route, many of the facilities planned in the 1950s and 1960s were realized and are still in operation. And more importantly, the new ideals of the optimal prison and the intense debate about the design of the industrial prison in the mid-twentieth century reintroduced on a new scale and in a new manner the issues and questions of communication that were important for the cell prisons of the nineteenth century. Communication was no longer mainly an "evil" (Meranze, 1996, p. 250) but instead was seen as an important dimension of life in prison that was facilitated actively through new communal areas and new possibilities for movement of and interaction between incarcerated individuals (figure 4.4) as well as between incarcerated individuals and staff. Communication and new communication systems were increasingly imagined as a solution—perhaps *the* solution—to the problems of incarceration.

In the older regime of the cell prison, the staff was explicitly forbidden to engage in conversations with incarcerated individuals (Mentzer, 1878, p. 26). In the industrial prison, conversations with staff were considered an

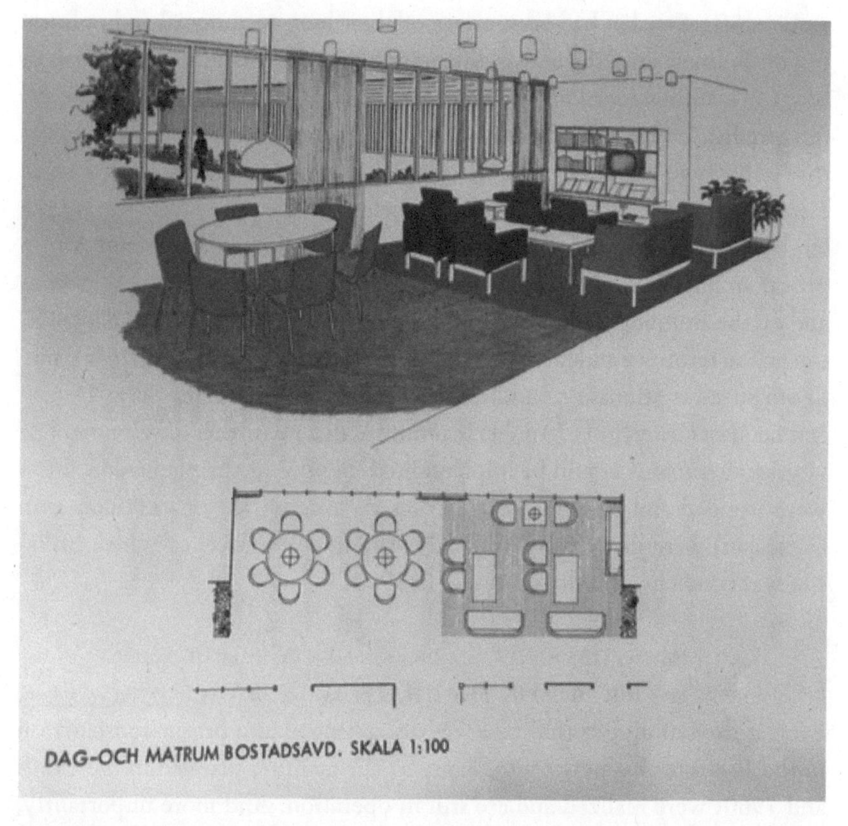

Figure 4.4
From Fångvårdens Byggnadskommittee, 1956–1972, *Slutrapport*, Stockholm 1972. This example design for a communal area in a Swedish prison facility invites communication and interaction between incarcerated individuals. Large windows open toward an outdoor area where two people are seen walking together. The room also includes a television set. Public domain.

essential part of the rehabilitation process. Trusting relationships and conversations were at the heart of the new therapeutic ideals of the postwar penal regime. The architecture and interior design of the industrial prison mirror these ideals through, for example, the new communal areas. When describing a smaller facility (for 150 people), the Building Committee, stated that

> the inmates are distributed in four houses, one low-security facility for 20 people and three-high security facilities for 40 people each, in concrete and with the new type of security windows developed by the committee. In the houses for 40 men, the entrance and guardroom is placed in the middle of the house, which creates two separated groups of cells. Each such group is further divided into two units, by placing the living room and kitchen in the middle. *This way, we get away from the sad corridor-character*. (Fångvårdens byggnadskommitté, 1956, p. 10, emphasis added)

Community areas, atria, and possibilities for freer movement for incarcerated persons were part of this strive for a new distribution of space within the prisons. The new ideal facilities were flattened, with only one-story houses and no basements, which created new forms of movement. The older prisons generally had several stories and were built as "castles," rising "unmistakable and daunting" (Luckhurst, 2019, p. 199) in the city centers, but the new prisons were their opposite: a distributed and scattered system of small buildings among and within which incarcerated moved themselves.

The new focus on communication in the postwar industrial prison was also mirrored in how technological solutions were incorporated in the building in order to enhance communication between staff and incarcerated individuals. One such feature was the central radio, installed in each cell. The radio was turned off at nighttime but worked around the clock as a transmitter-receiver, so that the incarcerated individuals could communicate with the staff in the guard room, and vice versa. The cells also had wall-mounted book shelfs. Books and magazines were among the media products that now entered the prisons. Beside these media, newspapers, gramophones, films, theater and later television were part of the media environment in Swedish institutions. This followed the new penal legislation from the 1940s that made access to culture and media a fundamental right for incarcerated individuals and hence marked a strong shift from the era of the cell prison where media and communications were almost entirely prohibited. Hence, the new penal regime also needed a new architecture to accommodate media within the prison. The movie theaters at the larger prison facilities are a clear-cut example of this shift. Other examples of how the new media environment is made possible through architecture and interior design are the wall-mounted radio systems in the cells, telephone booths, space for communal television, and libraries.

If the cell prison had an architecture that facilitated solitude and isolation, the industrial prisons of the 1960s, had an architecture that enhanced community and movement and was aligned with broader trends in the architectural community following the postwar years, which Reinhold Martin (2003) has documented in his study on the organizational complex. Martin shows how cybernetic ideas of decentralization, horizontal networks of communication and transportation lines, and the "application of redundant feedback loops" became a part of urban planning and architecture. Technocratic and aesthetic research, he argues, accelerated in directions of increasingly efficient mechanisms of self-regulation and self-organization as a response to the Cold War threat of a nuclear strike that would bring about the demise of centralized governmental authority. This mode of thinking spilled over from a military- and defense-oriented discourse and formed patterns of organization in other realms of architecture and social engineering of the time (Martin, 2003). This is evident in the new prison architecture and the way that space is distributed within these facilities: the small and to a large extent self-organizing community is the nucleus in these places; buildings are scattered and form a horizontal network in which the incarcerated can move somewhat unrestrained.

Nevertheless, the prison is still a prison: incarcerated individuals cannot be allowed to move too freely or communicate completely as they wish. Therefore, a range of technical systems that regulate and structure the movements and interactions within prison must be put in place, the surrounding walls being the most clear-cut example of such a restriction. But movement in the prison is also regulated through a system of underground transportation tunnels through which incarcerated individuals move, constantly watched by guards in a distant control room who open (or not) the electronically controlled doors and passages and hence participates in the control of the "self-regulated" patterns of movement within the facility. The system is described as such by the head architect Tor Bunner:

> The institution's various buildings are connected by a combined transport and service culverts. The latter part, which contains the institution's supply lines, for

> heating, hot water, and electricity, is separated from the transport culvert by a concrete wall for safety reasons. In the transport culvert, food is transported from the central kitchen to the residential pavilions, clean clothes, and dirty clothes to and from institutional storerooms, rubbish and waste from the pavilions. And of course, the staff and the inmates use this culvert to walk dry-shod between the different buildings. (Bunner, 1967, p. 28)

In this quote it sounds as if incarcerated individuals at the prison freely chose to use the culvert for convenience (to walk dry-shoed). However, the opposite was true. The culvert was the only way of transportation that incarcerated individuals could use in order to move around the prison facility; they had no possibility of choosing to walk outdoors. In that sense the architecture is not so anticorridoric as it might first appear, but the corridors are now underground, hidden from view, and take the shape of transportation tunnels and culverts that connect the different parts of the facilities. The precondition for this form of prison architecture was the incorporation of media technologies for surveillance into the structures of the building. The Swedish Prison and Probation Service's Building Committee wrote in a 1956 report that "if the ongoing tests with so-called industrial televisual surveillance turns out favorable, we intend to propose that the security is strengthened with television cameras, placed on the walls, in the workshops and in the corridors of the sleeping barracks. The television screens should be placed in the central guard room" (Fångvårdens byggnadskommitté, 1956, p. 10).

Without media technologies such as CCTV and other technological devices for monitoring movement and actions of incarcerated individuals, the industrial prison would be unthinkable. And therefore, such technologies were rapidly developed and implemented in the new facilities constructed in the 1960s.

The general belief in architecture as a key element in the reform and care of incarcerated individuals remains to this day in the Swedish Prison and Probation Service. A recent research report published by the agency claims that "Sweden is at the forefront internationally in terms of developing the concept of supportive prison environments" and that it is "it is worth underlining the value of architecture and design in enabling good relations, program activities and rehabilitation" (Grip et al., 2018, p. 40). The concept of "supportive environments" is somewhat vague. However, the central dimension of supportive environments is that they should enhance "communication" and "interaction" among the incarcerated individuals and between the incarcerated and the staff. One of the models for this new era of prison architecture is to be found in Halden Prison in Norway, which is often referred to in international discussions about new-generation prisons. Halden, opened in 2010, is "designed for maximum interaction and communication" (Hancock & Jewkes, 2011, p. 621). This resonates with the report from the Swedish Prison and Probation Service in which communication is one of the most prominent buzzwords and almost all "positive" features of architecture and interior design are acknowledged to enhance communication. Even connections that seem a bit far-fetched are included, such as the "dimmer lighting is preferred in certain contexts as it encourages communication" (Grip et al., 2018, p. 38). While contemporary discussions on prison architecture have moved quite far from the nineteenth-century "model prisons," prison architecture still circles around the problem of communication. If prison architecture, as argued by Hancock and Jewkes (2011), conveys messages about the individuals confined within them, their supposed characteristics and how they are expected to behave, then the twentieth-prison most certainly expects incarcerated individuals to communicate.

WATCHED BY THE WALLS: SURVEILLANCE IN THE INDUSTRIAL PRISON

One of the contemporary fantasies of a technologically enhanced prison is the so-called prison without guards. The most internationally renowned experiment of this kind is taking place in Singapore, where a range of advanced technological solutions are being deployed in order to replace the human employee within the prison with artificial intelligence and smart monitoring systems. Technologies such as video analytics systems, data analytics, biometric sensing, and facial recognition are hoped to replace human surveillance, while the use of smart technologies, such as tablets, apps, and digital kiosks are supposed to replace or at least simplify the treatment, care, and resocialization of incarcerated individuals (Khair, 2018). The attempt to creating a prison without guards echoes other contemporary attempts in automation in both public and private enterprise. It is also branded as a contemporary, modern, future-oriented idea that is to enhance effectiveness and rationality, freeing officers from guard duties to perform higher-order jobs.

The idea of automating the prison, however, is not so new and is closely linked to the built environment of the prison. This must be understood against the background of a general effort to reduce costs in the new prison development. One of the three main principles formulated in the government's proposition from 1950, which marked the start for new prison development, was that "it is necessary that the organization is designed so that the devices regarding buildings and equipment do not become more expensive and especially that the staff does not become larger than required for a sensible compliance with institutional pedagogical requirements and safety considerations" (Prop. 131/1950: Riktlinjer för fångvårdens framtida organisation, quoted in Gahrton, 1971).

In 1956 in the very first report issued by the Swedish Prison and Probation Service's Building Committee, "tele-technical surveillance" as a way of reducing the costs of labor and to automate the services provided by human guards was introduced for the first time. The technology was still experimental, but in general tele-technical surveillance is older than many conventional histories give the appearance of. The logic of CCTV and other technical surveillance systems has been traced to the early use of photography within criminal institutions and police departments that began in the nineteenth century. The ideas of physician Cesare Lombroso and his notion of a "physiognomics" reveal that the "criminal type" was one of the backgrounds to the development of photographic archives of criminals and prison photography. For Norris and Armstrong (1998), CCTV in the twentieth century was a continuation of this tradition while also relying on the "panoptic" ideal of the prison developed in the same period. By connecting these two technologies—the panopticon as an architecture of inspection and surveillance tool and photography as a technology of visual representation—the idea of a tele-technical surveillance system was born, they argue. Early experiments with CCTV and surveillance cameras were conducted in prisons and camps both in the Soviet Union and Nazi Germany already in the 1930s, but it was after World War II that the technology was developed more broadly and was implemented in both the prison and policing context in Western Europe and later in the US (Lauritsen & Feuerbach, 2015). As Chris A. Williams (2003) has shown, the police in Britain started using CCTV in the 1950s, but similar to how it was implemented in prisons, its use was developed and implemented within existing organizational frameworks. Already before actual CCTV there was a surveillance system in place, relying on other technologies such as the telegraph and the telephone for collecting surveillance data from public places and visually representing them, such as on a map in a centrally placed control room, in order to survey public places. Furthermore, as Williams (2003) shows, the implementation of CCTV by the police in the 1950s corresponded also to internal organizational demands of the police force as much as of changes in the surrounding world. There was at the time an increasing will to survey the police officers so that the managers would be able to survey and follow where they were (on patrol) and what they were doing.

This is equally applicable to the use of surveillance technology within the prison context. Here as well, the surveillance technologies introduced in the 1950s must be seen as extensions and further developments of the logics introduced in the prison in the nineteenth century. The "all-seeing eye," previously manifested through architectural principles that allowed for human surveillance, was enhanced and partly replaced with the camera lens.

Within the prison system, the surveillance technology introduced was a response to internal logics within the system of criminal care as such. Changes in the penal policy of 1945 that created a need for a new form of prison architecture (with factories, leisure areas, and greater freedom of movement for incarcerated) in turn needed new technologies for surveillance and control. The changes in penal policy also put higher demands on therapeutic approaches, care, and efforts for resocialization. In order to make such a shift within the prison system, the human efforts of guarding incarcerated individuals had to be transformed into human efforts for facilitating care and therapy. Communication, both among the incarcerated themselves in small groups and between the incarcerated and staff, is the new ideal, and a number of new professions—experts in communication such as psychologists, therapists, curators, and pedagogues—enter the prisons. There was a need to automate parts of the work in prisons, freeing officers from guard duties to perform "higher-order jobs," as expressed in the contemporary discourse. Tor Bunner, the head architect of the new prisons, expressed this view "the staff must be given possibilities to partake in the treatment [of the incarcerated individuals]. This is possible if we use the modern technologies for surveillance and safety in order to as far as possible relieve the staff from unproductive surveillance duties" (Bunner, 1967, p. 18).

This was not, however, uncontested, and in a book on prison surveillance written in the same period (1962) as a textbook for prison guards, the author nuances the more enthusiastic hopes for automation of surveillance in prison. "Technical equipment—television surveillance, alarm systems, high walls, high perimeters, and secure grids, etc.—is of good use at the closed institutions.

But without an alert, interested and knowledgeable staff, good results will not be achieved. Ultimately, it depends on the individual officer how the result will be. Personal effort of each man is required" (Ringius, 1962, p. 14).

The officials on the Building Committee were, however, more hopeful in regard to the new technology, and their suggestion was that regarding the issues of "security," the "aim should be that as far as possible [to] replace living labor with technology" (Fångvårdens byggnadskommitté, 1972, pp. 10–11). As they conclude, "to limit the costs for guards in benefit of therapeutic staff could only be done through facilities with high surrounding walls, different kinds of technical surveillance equipment and to direct a fair share of the traffic between different parts of the facility to underground culverts."

The Building Committee coordinated the piloting of tele-technical surveillance in a number of facilities beginning in 1956. After the piloting and proof of concept, they suggested introducing tele-technical surveillance across the whole country by placing cameras at the surrounding walls and perimeters and in the workshops and the corridors of the housing units. All facilities should furthermore coordinate the televisual surveillance from one central control room from which the cameras could be steered (Fångvårdens byggnadskommitée: Anstalter i Tidaholm, Norrtälje, Säve, Hälsingborg, 1956). In all their reports, the Prison and Probation Service's Building Committee makes very careful comparisons between the costs of every facility with and without tele-technical surveillance technologies, showing how these technologies contribute essentially to saving money. The foremost goal set by the Building Committee was to free guards from surveillance tasks:

> Taking into account that all security personnel should devote as much of their time as possible to care and not purely guarding tasks, the Committee considered that security technology substitute manpower to the greatest extent possible, especially for the reduction of working hours expected. . . . Time savings of guards in favor of care could only be achieved in a closed institution by the use of high perimeter walls together with various types of technical surveillance equipment and by shifting a large part of the traffic between the buildings of the establishment to culverts. (Fångvårdens byggnadskommitté, 1972, p. 11)

The first facility that was equipped with tele-technical surveillance was the prison facility in Norrtälje. Several news reports published in connection with the opening of the facility highlight CCTV technology. One article published in 1959, the year of the opening, describes the new surveillance infrastructure with fascination, stating that

> 16 cameras follow the prisoner at all times. TV surveillance of the walls and gates of the prison. 16 cameras monitor the 1100 m long wall, what the cameras see is shown on two TV screens, located in the central control room of the prison. An automatic switching device connects two cameras at the same time and scans each part of the wall. In the event of any suspicious movements in the terrain around the walls, the guard can contact other officers inside or outside the prison area using a portable radio telephone. The camera system is of the latest technology and will work with the best possible reproduction both under strong sunlight during the day and in the relatively weak night lighting. Furthermore, cameras can withstand temperature differences. Norrtälje prison is the first prison in the country to use television technology for control and surveillance purposes. Thanks to TV surveillance, fewer guards are needed. In fact, it saves a dozen men, which means a significant reduction in costs. (Arbetartidningen, 1959)

One of the former guards who worked at the prison from its opening in the late 1950s until his retirement in the 1990s described the tremendously different feel of the facility to us in an interview. Having previously worked at a penal colony—a remote labor camp—with access to electricity only during parts of the day. To him, the new facility in Norrtälje seemed like arriving in the future, like heaven on earth really, and the surveillance was perfect. The portable radio telephone allowed him to remain in touch with the central control room while doing his rounds around the yard and the wall (figure 4.5). The technology also became an integral part of the training of officers, teaching them how to use the radio telephones and navigate the cameras. In a special issue of the magazine *Byggnadsindustrin* from 1967 devoted to the new prisons being constructed in Sweden, the main architect of the building committee Tor Bunner writes a long article about the principles behind the new constructions. Two whole pages in the article are devoted to the "electronical and tele signal-facilities" at Kumla Prison. Not only the CCTV system but also lighting, alarm systems, internal communication systems, window and wall alarms, and a system of built-in microphones in the cell walls activated by sound and monitored by the central guard are discussed in detail. Bunner concludes that "the safety at the Kumla-prison rests essentially on the low-voltage installations reported here. Properly managed, the facility significantly increases security and gives staff an increased sense of safety" (Bunner, 1967b, p. 31).

Figure 4.5
Norrtälje was the first modern high-tech prison. Here the control panel in the central guardroom. 1956. Uppsala-Bild/Upplandsmuseet. License: Attribution-Non Commercial—No Derivatives (CC BY-NC-ND).

The technology was at the outset bought from companies on the open market, including the German company Siemens but also the Swedish company Securitas that in 1949 started a technical division. Also, publicly owned companies such as the Swedish company Telub delivered CCTV systems and other tele-technical surveillance equipment to the Swedish Prison and Probation Service. Telub is interesting, since it highlights the connection between military technology and development within law enforcement and prison tech. In 1964 the company began supplying the military with tele-technical equipment and became an important part of the military-industrial complex of the time (Kriminalvårdsstyrelsens arkiv, Vård- och tillsynsavdelningen 1969–1981, Diarieförda handlingar, vol. F1:76). Toward the end of the 1960s, however, the costs for maintenance and development of technologies for surveillance increased rapidly, and the need for an in-house solution became more pressing. In 1972 the Swedish Prison and Probation Service started its own "tele-technical workshop" (Bergsten, 1983) in which design, construction, and maintenance for surveillance and communication systems were handled.

The purpose of tele-technical surveillance in the early pilot projects was mainly focused on gathering evidence to prevent flight attempts and smuggling of contraband into the facility and were hence part of the overall safety and security strategy at prison facilities. Over the decades that technology also became increasingly important in preventing self-harm and drug use. Cameras were therefore installed not only around the perimeter walls and gates but also in communal areas such as dining and television rooms as well as the kitchen and the medical unit (Allard et al., 2006).

Clive Norris and Michael McCahill (2006) describe the status of CCTV in prisons as constituting an intersection of penal regimes combining parts of the postmodern penal regimes, including ideas of preemption and prediction, for example, with the help of video analytics and the modern system still being partly intact. CCTV allows for "habituated anticipatory conformity" of incarcerated people that are under the impression of being constantly watched by not only the human eye but also smart devices that are used to predict future behavior. However, CCTV is a sociotechnical system that does not have criminological relevance in and of itself. Only by being embedded in organizational practices and systems through adjusting surveillance routines to the technological infrastructure as well as integrating technological education into the general training does CCTV receive social relevance and bearing.

Importantly, and as mentioned above, CCTV is also a prerequisite for prison architecture. As incarcerated individuals move around the perimeters and through the underground culvert system, their movements are registered by cameras and can be monitored and managed from the central control room in which guards, from a distance, open and close doors and control the flow of individuals through the facility so that only the "right" forms of contact between incarcerated individuals occurs.

PRISON MEDIA ARCHITECTURE: FROM THE CELL SYSTEM TO THE (POST)INDUSTRIAL PRISON

The main protagonist of *Grundbulten* describes the architecture of Hall Prison as one organism; one part is functionally connected with the other, and all parts rely on each other to work. Prison architecture is a closed system that organizes bodies and objects in time and space through the way in which communication is allowed and constrained in corridors, workshops, tunnel systems, control rooms, and cells. Prison architecture is a medium for the different penal regimes—the ideas and ideologies of crime and punishment—that have developed since the nineteenth century. The investigations, reports, drawings, and sketches express sociotechnical imaginaries of the role of technologies and mediation in the act of punishment and reform of incarcerated individuals and ultimately mirror as well as configure ideas of subjectivity (what is an individual) and of sociality (what is society). As such, the history of prison architecture is also a good example of the changing and contested nature of such imaginaries. While some of the ideas of the many architects of the system never left the drawing board, many others did, and the imaginaries materialized and crystallized in physical structures and penal practices. In this process, the lofty hopes for "automation," "humanization," and "effectiveness" often were tamed or at least had to be adjusted to the realities of everyday life in prison and of the agency and resistance of the individuals living their lives in prison. While ideals and imaginaries of prison architecture have changed, some of the themes and threads in prison architecture seem stable and recurrent: the tensions between surveillance and rehabilitation, control and community, and visibility and invisibility and the recurring problem of communication.

The architecture and design of the modern prison are highly intertwined with and connected to communication and media technologies and hence constitute what we call prison media architecture. While Beatriz Colomina (1994) argues that architecture only became truly modern through its relation to mass media in the late nineteenth and early twentieth centuries, we argue that architecture itself is a medium that constrains, channels, structures, and enables communication. For Colomina, thinking about architecture means "to pass back and forth between the question of space and the question of representation. Indeed, it will be necessary to think of architecture as a system of representation, or rather a series of systems of representation" (p. 13). However, architecture is also, as we have seen, an infrastructure of communication, and the problem of communication has been a key question for all prison architecture styles. From the outset, the problem concerned mainly how to prohibit and prevent communication while at the same time surveying the incarcerated individuals. Hence, the main solution meant creating architecture through which incarcerated individuals could be seen, heard, and registered at all times. Gradually the focus changed to creating an architecture that enhances and creates communication both as movement and as interaction for the incarcerated individuals. The culverts and CCTV system of the high-modernist industrial prison was in this respect an architectural—and communicative—innovation that also underlines the symbiosis between architecture and media technologies.

5

IMAGINING THE PRISON: PRISON MEDIA TECHNOLOGIES

The novel *Grundbulten* is a collection of imaginaries of the prison: of incarcerated individuals, the everyday life within the prison, the quarrels, the exchanges with guards, and the longing for the world outside the walls. Large parts of the narrative of *Grundbulten* also refer to the role of technologies and the affective dimensions of constantly living with automated surveillance, adding another layer of humiliation.

> It is believed that there was an intimate collaboration between business and government. Namely, the government is interested in getting an experimental model for the suburbs of the eighties and nineties, the electronically monitored ghettos made of concrete. According to that school of thought, the result has exceeded expectations for both business and government. Because it has been shown that you get used to such things as constant electronic surveillance, that surveillance cameras are everywhere watching you, that all doors are closed and opened electronically by someone you never see, that in the ceilings, in culverts and corridors there are microphones that can record all conversations, that inside the cells the radio acts as a receiver. You do not dare to say anything and maybe that is in fact what the idea is, not to listen to what people say but to make them shut up, go for themselves or just bullshit. (Ahl, 1974, p. 25)

Besides the narrative of humiliation, the quote above also hints at another prominent but overlooked aspect of prison media, namely their test bed character. The idea of prisons as test beds for novel technologies was already pointed out by Michel Foucault when he argued that "the Panopticon was also a laboratory; it could be used as a machine to carry out experiments, to alter behaviour, to train or correct individuals. To experiment with medicines and monitor their effects. To try out different punishments on incarcerated individuals, according to their crimes and character, and to seek the most effective ones" (Foucault, 1979, p. 203). The prison is imagined in Foucault's terms as not only a site for distinguishing between the good and the evil, a site of discipline, but also a place of experimentation for disciplining technologies outside the prison walls. Of course, Foucault thought of technologies as specific practices of controlling and disciplining populations rather than technological apparatuses. As institutions for disciplining, prisons are then not only material sites of punishment but also expressions of social imaginaries, in a broader social sense capturing shared expectations in a society and their underlying norms and in a more specific sociotechnical sense envisioning underlying infrastructures as "collectively imagined forms of social life and social order reflected in design and fulfilment of nation-specific scientific and/or technological projects" (Jasanoff & Kim, 2009, p. 120). Societal institutions and actors such as courts, the media, and policy makers have the power to underline certain sociotechnical imaginaries while sidelining others. For example, strategy documents of the Swedish government highlighting the potential and importance of artificial intelligence to solve crucial societal challenges, including crime and its punishment, are one step in materializing sociotechnical imaginaries. These imaginaries not only relate to specific ideas about technologies but also encompass normative ideas about how the good life and society ought to be. In that sense, they express and reify a shared sense of good and evil (Jasanoff, 2015). Hence, sociotechnical imaginaries carry normative aspects while always remaining provisional, imperfect, and under construction (Willim, 2017). This means that they are not merely fantasies but instead are translated into specific projects and practices, such as the building of larger digital infrastructures at prison facilities. Sociotechnical imaginaries constitute what Ingrid Forsler (2020) has called soft infrastructures that govern. As policies, they structure and guide political projects such as digitalization of corrections that have implications for hard infrastructures such as prison buildings. However, not all future visions and ideas are automatically sociotechnical imaginaries. Astrid Mager and Christian Katzenbach (2021) argue that they need some kind of collective resonance that has implications for the allocation of resources as well as translation into practices that contribute to their institutionalization. In that sense, sociotechnical imaginaries move beyond merely discursive dimensions.

Hence, we have to combine the focus on sociotechnical imaginaries of the future with their concrete implementation of testing technologies that move in and out of the prison. One of the starting points here is that prisons are intricately linked with technological development, although this linkage often remains under the radar of public attention. Similarly, the military has earlier been identified as a major driver for technological development especially of communication technologies with the Advanced Research Projects Agency and later the Defense Advanced Research Projects Agency investing heavily in emerging technologies and launching the Advanced Research Projects Agency Network, commonly known as the ARPANET, as the network of networks in 1966 and predecessor of the internet (see, e.g., Brunton, 2013).

While the contribution of the military complex to the development of communication technologies has been broadly acknowledged, this is not the case for the prison context. Here, it is a rather a picture of absence of technologies or a melting pot of obsolete technologies such as cassette tapes that is drawn when talking about places of incarceration; digital media technologies in particular seem evacuated from the prison space. This picture is enhanced by reoccurring stories about the digital illiteracy of incarcerated individuals with long-term sentences being either completely denied access or overcharged for using tablets, including sky-high rates for reading and sending emails as well as accessing pictures (Järveläinen & Rantanen, 2020; Wang, 2018). Even less so, prisons are connected with cutting-edge technology development. However, quite the opposite is the case, and there are a number of examples that illustrate the connections that prisons have with technological innovation. The prison of the nineteenth century gave birth to the treadmill that is now standard in gyms around the world (Doan, 2015; Peters, 1999), ankle monitors are remediated as wrist-worn trackers (Bernard, 2019), and early approaches to gamify and gratify positive behavior have been tested in the prison context (Hibbert, 1963). In more contemporary terms, correction facilities are increasingly implementing the latest digital technologies especially for administrative and surveillance purposes (Jewkes & Reisdorf, 2016; Reisdorf & Jewkes, 2016).

PRISON TECHNOLOGIES

In this context, we need to consider the specificities of technologies that emerge in the prison context, technologies that we call "prison tech." Prison tech is developed in particular for the prison context but has the potential to transition into other use areas and the general consumer market. In the Swedish context, the entanglement between technology development and punishment can be illustrated with one of the central publications from within the corrections systems authored by Torsten Eriksson, the director general of the Swedish Prison and Probation Service between 1960 and 1970. Based on his experiences as a civil servant (assistant and later actuary) in the second half of the 1930s, he wrote several programmatic books mapping the state of the criminal justice system. In his first publication, *Crime and Society* (*Brott och samhälle*) from 1939, he dedicates one chapter to the analysis of the Swedish penal system, establishing links to other jurisdictions primarily Germany, the United States, and the United Kingdom. This specific chapter evaluates the state of the art of the criminal justice system ranging from the classification of criminal behavior and forms of punishment to moving from corporal punishment increasingly toward prison sentences instead. Four major questions drive his investigation forward, namely what acts should be criminalized in society, how should society organize crime prevention, how should punishment be organized, and how can crime be prevented. Eriksson's main starting point is that punishment that cannot be practically implemented forestalls the criminalization of this behavior. In that sense, he argues that what is criminalized is not only a question of what is considered harmful for society but also how if it can be punished. He takes Prohibition in the US as an example. Not only had it been difficult to punish criminal behavior according to Prohibition regulation, but more importantly Prohibition led to the emergence of well-connected criminal networks. For the Swedish context, Eriksson mentions illegal abortions that because of the sheer number can hardly be prosecuted and punished to the full extent. This would lead to thousands of women behind bars annually, he argued. But it is not only the mere possibility (or impossibility) to prosecute certain forms of harmful behavior. Part of the argument is also the increased technological finesse developed for police investigations and prosecutorial work that go into proving crime and identifying perpetrators. The emerging technologies for policing and punishing of the 1930s are documented in numerous images in Eriksson's book.

The images in Eriksson's (1939) book include novel control rooms of Scotland Yard employing radio technology for investigating and surveillance of organized crime (figure 5.1) as well as the use of radio transmitters integrated into police helmets of the British police. Besides technologies for surveillance, Eriksson also refers to technologies to register and identify criminals including the photographing and filming of suspects (figure 5.2). Registering criminals and the incarcerated with the help of photos was one

Figure 5.1
Control room of Scotland Yard in Eriksson (1939, p. 233).

Figure 5.2
"The latest identification method—sound film—is beginning to be used" (Eriksson, 1939, p. 253).

of the earliest applications of photography. Already in the 1850s most (male) incarcerated persons were photographed, as images were widely considered superior to the written descriptions that previously had been used to keep registers on incarcerated individuals. The handbook for prison directors in Sweden, published and distributed among prison wardens in 1878, has a long and detailed instruction on prison photography stating, among other things, that "all male prisoners . . . who are considered dangerous or unreliable for public safety, shall be photographed" (Mentzer, 1878, p. 48). Copies of the photographs with written descriptions of the depicted individual would then be sent to police authorities in all major cities.

A number of emerging technologies are also used in the prison context, as Eriksson's (1939) book demonstrates. He engages, for example, with a novel mirror system that allows for constant surveillance of the incarcerated in visitation rooms. Eriksson's mapping of police and prison technologies from the 1930s makes clear that technological development and testing is all but new in this context. It is also a testimony to the fascination of the correction sector with latest technologies that is imagined to either help solve crimes or support the authorities in their approaches to punishment.

TECHNOLOGICAL MEDIATORS FOR PRISON TECH

At contemporary security and prison technology expos, we encountered similar established and emerging technologies for surveillance, control, and punishment that Torsten Eriksson was fascinated with in the 1930s. Although the expos and technology fairs overemphasize the revolutionary and disruptive character of contemporary prison technology, the technological development in prisons is part of a longer history of technology for surveillance and control of marginalized populations. These technologies form a continuum ranging from targeting specific, often marginalized populations through predictive policing to large-scale broadly used technologies such as state surveillance of citizens. In that context, Shoshana Magnet (2011) explores the failures of biometric technologies being first developed in the prison context to later being transferred into the welfare sector and regimes of border control. She shows how vulnerable groups become objects of body fetishism aiming to develop objective measures for the classification of people. Often these groups are coerced test subjects for new technologies that would provoke resistance in other use areas. In the context of developing biometric technologies in the prison, Magnet argues that

> prisoners' bodies are valuable commodities to biometric companies, providing the industry with a captive test population for assessing the efficacy of these new identification technologies. While the banking industry might be cautious about introducing biometric technology to their clients for fear of scaring off potential customers, prisons do not have the same restrictions. In this way biometric technologies are key to the intensification of particular forms of monitoring, as they provide a way for companies selling surveillance technologies to try out products that would be rejected by the general public. (Magnet, 2011, p. 63)

In a similar manner, Brian Jefferson (2020) highlights the ways in which the state has contributed to technological development through investing extensively in digitizing punishment and policing. Both Jefferson and Magnet (2011) show that the entanglement of technology and punishment or policing goes much further back than currently emerging forms of digital carceral technologies, as we have also seen in the technological interest of Torsten Eriksson in the 1930s–1940s in Sweden. Magnet and Jefferson demonstrate how these technologies are connected to forms of racialization. Similarly, Simone Browne (2015) disentangles the connection between the historical formation of race and contemporary surveillance technologies. Ruha Benjamin (2019b) turns this argument on its head in the edited volume *Captivating Technology*, which gathers contributions that explore how the penal logic is dispersed into other institutions and social spheres through the export of technologies—both digital and analog—that have initially been developed for surveillance and punishment. The various contributions show how the "sticky web of carcerality extends even further, into the everyday lives of those who are purportedly free, wrapping around hospitals, schools, banks, social service agencies, humanitarian organizations, shopping malls, and the digital service economy" (Benjamin, 2019b, p. 2). While exploring the general dispersion of captivating technology in society, the authors focus on discriminatory design and hence argue that there is a paradox of techno fixes that are supposedly developed and implemented to overcome human bias while being based on biased and discriminatory logics with different implications for poor and racialized people than majority groups. Benjamin (2019a) explores the linkage between digital technologies and racial discrimination further in her book *Race After Technology*. Arguing that the seemingly neutral technologies establish a "New Jim Code" with reference to the Jim Crow laws establishing racial segregation in the US in the late nineteenth and early twentieth centuries, she explores how people are implicated differently when it comes to technologies.

In order to extend these earlier explorations of carceral and surveillance technologies, we have attended trade shows and technology expos and consider them as platforms for technological mediators of prison tech. These technological mediators constitute actors of tech development and innovation that are often dispersed in different sectors but convene at tech expos, which consequently emerge as important sites of technological advancement while not being directly involved in the research and development process. Here we link to earlier studies of tech interlocutors that were explored by Joshua Greenberg (2008) focusing on neighborhood video rental stores that were constitutive of the VHS and film industry but rarely considered as such. Similarly, Christina Dunbar-Hester (2014) has explored low-power radio activists as advocates for and developers of technology. Li Cornfeld (2020) explores sex toys and the porn industry as areas of early testing and implementing of new technology. She argues that "media and technology of sex on the technological vanguard [was] born either out of an essential human desire for sexual expression or market logics that incentivize producers of illicit goods to engage new means of production and distribution" (p. 96). These studies show how diverse social groups contribute to technological development in varied and often unexpected ways.

Based on observations made at the trade shows, prison tech can be roughly divided into three main categories: surveillance technologies (including contraband detection and CCTV solutions), communication (including tablets, email, and telephony), and e-learning (including apps, platforms, and online libraries). While surveillance technologies and communication dominated in terms of variety and number of vendors, the number of e-learning solutions was considerably smaller. The companies focusing on prison technologies can be distinguished further into prison tech insiders, prison tech specialists, and tech generalists.

PRISON TECH INSIDERS

Prison tech insiders are companies with a narrow portfolio focusing on one specific tech solution that was often developed within one facility by (former) correction officers to meet specific needs of the facility. An example of this type of prison technology company is Tech Friends, Inc. Founded by a former correction officer, the company offers different technologies including telephone kiosks, debit calling, and video visitation solutions. At one expo, the founder and owner of the company was showcasing tablets with individualized and personalized applications. The tablet was custom-made and developed in-house in one specific facility first. When the developer retired as a corrections officer, he started his own company and now employs several coders developing his software. Tech Friends offers a tablet with email access and video calls for incarcerated individuals. Based on the data collected, the facilities are provided with automated content analysis of emails and messages exchanged through the tablet, preset searches for words alerting officers automatically, network analysis of individual incarcerated persons including who they communicate with and who they receive money from outside of the facility, and visualizations of network maps including information about how strong the links to other actors in the network are and what kind of relationship they have to each other. The company is currently seeking to expand and aims to sell its product to an increasing number of facilities.

Another example of prison tech insiders are in-house units that maintain and repair the tech infrastructure of the prison. In the Swedish context, a dedicated department for technological maintenance—the so-called Televerkstan—was first set up in the late 1960s when the scope of technological infrastructures in the newly built prisons including Österåker and Kumla expanded. This unit was responsible for maintaining and repairing surveillance technologies including CCTV as well as the telephone network. The unit also developed its own components and a full-fledged central radio system that was installed in all cells in the newly built prisons (figure 5.3). One article in the staff magazine *Kriminalvården* said half-jokingly about the radio system, "So brace yourself, Sony, Hutachi or whatever you are called, here we might have a real future competitor" (*Kriminalvården*, 1983). Despite the depiction of the unit as saving extensive costs for external services (*Kriminalvården*, 1983), it was never prioritized in the overall budget of the Prison and Probation Service and remained rather small in terms of staff numbers. Nevertheless, it persisted as part of the central administration in 1970s and 1980s, with a mobile service that served all facilities across Sweden. The unit has been rebranded as Krim:Tech, with its own trainee recruitment program for coders and the aim of developing its own smart solutions for

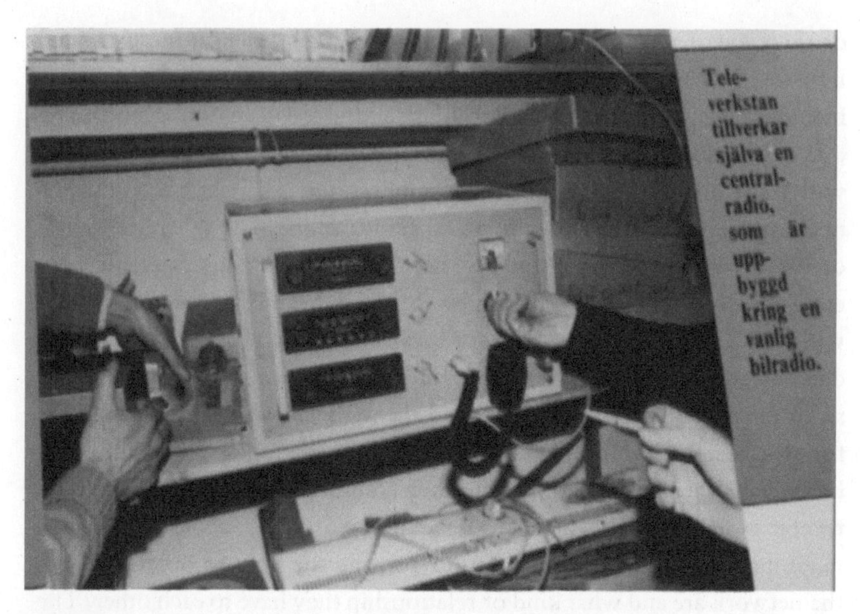

Figure 5.3
Workshop of the tech unit within the Swedish Prison and Probation Service in 1983. The caption reads "The tech unit itself develops a central radio unit that is built around an ordinary radio." *Source*: *Kriminalvården* (1983).

the Prison and Probation Service, including the implementation of test beds for radical innovation.

PRISON TECH SPECIALISTS

Prison tech specialists are larger companies dominating the correctional technology sector. They aim to maintain or further expand their markets. Two of the most prominent examples in the US market are GTL communications and Guardian RFID (Kaun & Stiernstedt, 2020b). GTL communications is one of the larger vendors that has been dominating the US prison technology market when it comes to communication and is increasingly integrating all three categories of prison tech (surveillance, communication, and e-learning) through the custom-made tablet solutions it has offered since 2015. The company now serves more than 80,000 tablets across the US in at least six states. GTL communication charges incarcerated individuals 5 cents per minute for content access beyond the base provision. Extra services include access to family pictures (50 cents per attached photo and 1 dollar per video attachment), video calls (25 cents per minute), chat

(25 cents per written message), and radio and streaming as well as library resources. In comparison, incarcerated individuals earn between 4 and 58 cents per hour depending on their assignment (Sawyer, 2017).

Attenti and SuperCom are two additional major prison tech specialists providing offender monitoring solutions for the corrections sector as well as private homes. Both companies supply ankle monitors for the corrections sector and are among the largest private contractors for the Swedish Prison and Probation Service. Attenti quantifies its operations by including the number of individuals being monitored each year.

TECH GENERALISTS

Tech generalists within the prison tech market are large international corporations that offer specialized solutions but have a broad portfolio beyond the corrections sector. One of the most prominent examples is the software package Offender360 by DXC Technology, a partner company of Microsoft. Offender360 is built on the Microsoft Dynamics 365 platform and is mainly used for jail administration, including record and assessment management. Reports on all records, filter functions, questionnaires, etc. can be processed through the platform, and risk scores dependent on assessment can be generated.

Many of the larger companies aim at an international market presence. Accordingly, they must adjust to national jurisdictions while making this change part of their selling point. For example, the German company Telio—prison tech all-rounder—describes itself as number one player in the European prison industry with a presence in 18 countries while ensuring "full compliance with local laws and regulations. Cultural complexity is managed through local teams in each country."[1] This is one illustration of the transnational character of prison tech that needs to adapt to specific national regulatory frameworks and distinct ways in which the criminal justice system is organized while maintaining the core functionalities of its products.

TECHNOLOGICAL BACKWARDNESS OF THE PRISON

Besides representatives from technology companies, the expos gather other actors of technological mediation in the prison sector. These other actors include representatives of the criminal justice system such as administrative officers, information technology officers, and management as well as experts including academic and nonacademic researchers and consultants. They are technological mediators co-constructing prison tech. While all actors listed are directly involved in the coconstruction process including the production of sociotechnical imaginaries, incarcerated people become coerced or indirect participants; talked about and imagined as objects to prison tech, they of course never participate in the expos or fairs. The involved actors produce not only the hardware and software of prison tech but also very particular sociotechnical imaginaries that legitimate the implementation, development, and use of specific prison technologies.

The prime starting point for imagining and developing prison technologies is the construction of a fundamental backwardness: technological development needs the idea of backwardness to be overcome and problems to be solved. In this context, prisons and the corrections sector more generally are often imagined as particularly lagging behind technological development. During one of the conference presentations we observed, a practitioner within the corrections sector emphasized that "in the prisons sector we are hopelessly behind. At least twenty years!" This mantra was repeated to us consistently by the prison officials. Digital technologies and digital development are in that context of backwardness imagined to not only improve small parts of the administration and prison experience but also radically reform how the corrections sector is organized.

Peter van de Sande, president of the International Corrections and Prisons Association, opened the conference "Technologies in Correction" with a paper tellingly titled "Digital Transformations":

> Before the digital revolution we had to buy vinyl, we had to buy the whole album. Now, I carry all the music in the world around in my phone, just listening to what I want to here. This has made my life so much easier. In a similar way, corrections could be redone if digital transformations would be embraced more openly. Digital technologies can revolutionize corrections to make our lives better and easier, but there is a general reluctance to technology, because staff is suspicious and in fear of being replaced. But they will to the very contrary be freed from monotonous tasks and can finally focus on their real work including rehabilitation and client contact. Digital transformation will contribute new ways "for exercising our business" and the corrections sector should be ready to be "digitally transformed."

There are different ways in which this digital transformation of the corrections sector is envisioned. During the same conference, an armchair discussion focused on the chief information officer perspective on digital transformations that included Simon Bonk of the Correctional Services Canada; Håkan Klarin of Kriminalvården Sweden; Teck En Loh, director general of the Singapore Prison Services; and Russel Nichols of the California Department of Corrections. During the panel Klarin emphasized that cross-sectorial partnerships—what he called the triple helix of academia, market partners, and the Prison and Probation Service—are the way forward. He stressed that opening the way for private partnerships would speed up the digitalization process. The corrections sector is part of a larger digital ecosystem and should acknowledge this fact by developing new models for collaboration. Klarin, for instance, imagined open-source sharing of digital solutions across borders, using a Swedish application developed for clients on probation—the *utsiktsapp* (Kaun & Stiernstedt, 2020a)—as an example. An important part of the panel discussion was dedicated to the question of why innovation takes time within corrections in comparison to other sectors. Here, the speakers identified three important aspects. First, there is a specific culture in corrections that forestalls change. The conservatism toward technological innovation is but one example of this culture. Second and connected, the technological innovation envisioned by the chief information officer panel presupposes a complete change in the ways in which corrections are managed, and this needs time. Large institutions are moving and changing slowly. Therefore, there is a strong need to constantly renegotiate the ideas about and attitudes toward technological innovation. Third, the corrections sector as a public endeavor competes for resources, and the introduction of new technologies is cost-intensive. Building a business case for public-private partnerships takes a very long time during which the tech industry has already moved on and potentially has lost interest in the collaboration. Nevertheless, the corrections sector is the right field for tech. The argument goes that technology can be an important tool to improve communities and provide possibilities for connection. Russel Nichols suggested during the panel that in the US, for example, access to email has changed the life for many offenders and their families in very positive ways. The strong emphasis on technological backwardness of the corrections sector calls for innovative ways of partnering up with the industry, as the arguments exchanged during the panel discussion show. It is the intricate link of the need to catch up while being in a vulnerable underdog position that makes unorthodox collaborations necessary and possible.

Despite or rather because of the established discourse of technological backwardness, the corrections sector provides a fertile environment for testing technological solutions and further developing applications to be used in other commercial contexts. A sales and marketing representative from an Australian company focusing on secure cloud solutions whom we met during one of the conferences stressed that many companies that are active within the corrections sector dedicate only a small part of their operations to prison services. Often, companies explore collaborations with the corrections sector while having a broader portfolio and catering to other sectors. His company, for example, integrates in existing CCTV systems artificial intelligence that detects unusual behavior. The company's services are currently piloted in several Australian prisons but could also be used for surveillance of other critical infrastructure. He also emphasized that the corrections sector is otherwise a very important line because of privacy concerns that would hinder introduction of certain technologies in other areas. The representative uses the example of integrating artificial intelligence–enabled CCTV analysis with monitoring of facial expressions and heart rate monitoring to detect stress and unusual activity prior to violent behavior. He personally does not like the idea that his heart rate is monitored whenever he enters a shopping mall, for example, while in prisons this technology might save lives. At the same time, the knowledge based on introducing certain applications in the corrections sector can lead to specific adaptations. The representative underlined further that technological innovation in the prison context always builds on existing infrastructures. There are rarely moments of innovative disruption or changes in the whole infrastructure. Usually, older systems are upgraded with new technologies that are built on exiting systems and technologies. Disruptions on a larger scale are anticipated for new prison buildings, where architecture is integrated with the latest technologies. This discourse of technological backwardness serves as context and necessary precondition for the innovative work and test bed activities at prisons.

DECENTRALIZING THE POSTINDUSTRIAL PRISON

THE DREAM OF TAKING THE PRISON HOME

One specific device that was developed in the corrections context and has been shown to have similarities with broadly used everyday devices is the ankle monitor. Andreas Bernard has previously argued that there are clear similarities between the technology of ankle monitors and contemporary devices for self-tracking including wrist-worn trackers and smart watches. He writes that "regarding the genealogy of certain conceptions of humanity in digital culture, however, it is significant that the location technology was first implemented to track individual people, and, like the historical development of the profile, that it was based on a contribution of applied psychology to the field of law enforcement" (Bernard, 2019, p. 45). The aim of early body-worn tracking devices was to control and correct the behavior of unstable or deviant individuals by technical means. Later this aim was adapted to serve self-improvement and nudging toward healthier lifestyles based on self-knowledge.

Looking closer at the ankle monitor, we are also following the idea of moving punishment out of the confines of the prison walls and into the homes of offenders and society more broadly. Following the logic of technological solutionism (Morozov, 2013), ankle monitoring, although not uncontroversial, has been proposed as the solution to the crisis of the prison. Combining different surveillance technologies, Bagaric, Hunter, and Wolf (2018) suggest that 95 percent of US prisons could be closed. The ankle monitor is but one very concrete sign of the dissolution of punitive logic into the social, constituting a form of mobile incarceration. Ankle monitors constitute the most obvious diffusion of the prison into society, disconnecting punishment from specific architectural sites and moving it into the home.

While Andreas Bernard (2019) mentions the inspiration of a Spiderman comic strip in the late 1970s for one of the first American judges to order ankle monitoring in a sentence, the initial developer of one of the first wearable monitoring systems, Schwitzelgebel (later taking the name Gable), states that he was inspired by the film *West Side Story* (Gable & Gable, 2016). Ralph Schwitzelgebel imagined a communication system that would prevent the murder of the main protagonist. Modifying a missile-tracking device, Gable designed several different portable receivers coupled with a stationary radio-frequency relay station. The initial tracking was in that way possible in real time and over five square blocks. The idea was to extend preventive counseling that otherwise was very limited geographically. The feeling of being observed and tracked should nudge the subject into lawful behavior but also contribute to a feeling of security and support. Tests with offenders ranged from monitoring over the course of a couple of hours to full-day surveillance. The first 16 participants in the experimental phase showed modest changes in their criminal behavior in comparison to the control group. In the following years Gable and his brother—who also constructed a low-powered FCC-licensed radio—refined the tracking technology, developing the two-way tactile belt that was part of a communication system for probationers. Through the belt the probationer could receive and send short messages—which Gable ironically compared with Tweets—when, for example, in a classroom or on a job site (Gable & Gable, 2016). The messages were intended as a form of positive attention and reinforcement. Gable was later involved in additional technological developments for the corrections sector including telemetric physiological characteristics—at least one of which is now integral part of fitness trackers—such as heart rate and galvanic skin response of offenders measured in natural social settings and compared with self-reported emotions of the subjects when they experience anxiety.

In 1994 Sweden was one of the first countries in the world to introduce ankle monitoring and with very little public scrutiny and critique (Nellis & Bungerfeldt, 2013). This was a time when the number of incarcerated increased exponentially due to major change in the laws, including the extension of criminalization of petty crimes now being increasingly punished with prison sentences as well as the abolition of the possibility to get the prison sentence turned into probation after half of the time has been served. These changes of course increased the costs for the Prison and Probation Service, and in 1993 it was estimated that at least 1,000 new prison places were needed to be able to handle the increased number of incarcerated.[2] As Mike Nellis has argued in the context of introducing electronic monitoring in England and Wales, there are very particular tropes of motivating and legitimating the introduction of technologies in the corrections sector. He argues that "it can be said that the electronic monitoring (in the broadest sense) is primarily grounded in, and expressive of, the managerial-surveillant enterprise—it facilitates the efficient, fine-grained regulation of human action in time and space" (Nellis, 2005, p. 180).

Swedish newspapers referred to ankle monitors by that time half-jokingly as an electronic leash. The first newspaper reports about ankle monitoring in the US appeared in Sweden a bit before their actual introduction in the Swedish criminal justice system. In 1992, one of the major dailies included a report about a man who was sentenced to surveillance with the help of an ankle monitor for 16 months.[3] The first experiments in Swedish pilot cities, among others Malmö, were accompanied with frequent media reporting as well as descriptions of the technology and human-interest stories on the experience with ankle monitoring by offenders. In one of the earliest articles from 1993 published in *Göteborgsposten*, two critical points are offered. First, the cost efficiency is a myth. Countering the main argument for ankle monitoring, namely the savings in cost, the article argues that ankle monitoring requires surveillance staff around the clock in addition to resources in the case of flight attempts. The second critical point being made is related to the question of who will be able to apply for ankle monitoring. The specific crime and length of the sentence play a role, but so do the preconditions of the offender. Not only do offenders need to live under stable conditions with their own housing and regular employment, but they also must pay a specific daily allowance for the device—in the 1990s around 50 kronas (about 5 euros) per day that was to be paid in advance and was donated directly to the Swedish Crime Victim Fund.[4] This leads potentially to a two-tier system of punishment that allows a privileged group to serve the sentence at home while the rest will be gathered in the prisons.[5]

In 2021, the conditions of ankle monitoring and serving a sentence at home include a curfew, responsibility for the equipment, a fee, and inspection visits. Offenders with ankle monitors are allowed to move outside of the home only for work, educational training, and therapy that is scheduled by the probation officer. The remaining time offenders are on curfew. In case the offender does not follow the preset schedule without explanation, the ankle monitoring will be turned into a prison sentence. The offender carries the sole responsibility for the technological equipment. The ankle monitor itself is robust and water-resistant and is connected via an application with the smartphone of the offender or a mobile receiver unit that is installed in the home. If the equipment is destroyed, the offender must replace it at his or her own costs. Probation officers conduct announced and unannounced inspection visits to check the curfew and no-drugs regulations.

In the Swedish online forum Flashback, users—both with and without experience from ankle monitoring—are speculating about the regularity of and schedules for unannounced inspection visits. There is, for example, the rumor that the Prison and Probation Service rarely does inspections after 11 p.m. In the forum, offenders also share emotional experiences of being monitored at home. Some argue that being monitored at home is harder and that the schedule is stricter than being in prison. You must stick to your time regime very closely, and the smallest violation is punished. It is also argued that the equipment sometimes has technical issues, causing false alarms in the control room and leading to nightly scenes including many police officers alerting the neighbors. Other offenders share that the experience of being monitored constantly at home in their familiar environment is hard to handle. In prison there are known areas that are largely unsupervised, but with an ankle monitor you are under constant supervision.

The monitoring system itself includes a radio sender with a short reach that is attached to the arm or foot and a receiver that is placed in the home (figure 5.4). If the offender moves too far from the receiver, a central computer is automatically alerted (figure 5.5). The offenders must keep a strict time schedule leaving for and returning from work, with only small window of a couple of minutes to move within as they are only connected to the receiver at home and are not tracked in real time while moving outside of the home. After a test period of two to three years, ankle monitoring was introduced on a larger scale after a governmental evaluation.[6] In 1999, the Swedish National Council for Crime Prevention published a report about the possibilities of using ankle monitoring for restraining orders after domestic violence, sexual

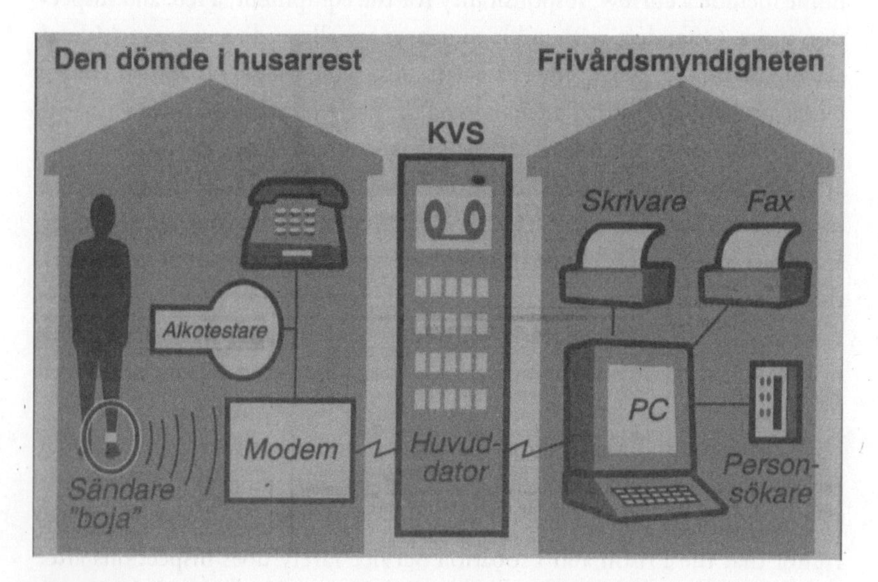

Figure 5.4
Ankle monitoring system of the 1990s in Sweden. *Aftonbladet*, May 12, 1994.

offense, and reoccurring stalking. The idea was to use the same monitoring system but that the receiver was to be placed in the home of the offended person and would alert the central control room in case the offender entered the radius of approximately 150 meters (Swedish National Council for Crime Prevention, 1999). However, these plans were never realized.

Today there are around 1,600 offenders who are under so-called intensive surveillance with electronic monitoring. Sixteen percent of the offenders are women, while 84 percent are men. The daily fees are now around 80 kronas (about 8 euros), with a maximum amount of 9,600 kronas (about 960 euros) for the whole monitoring period to be paid to the Swedish Crime Victim Fund (2014:1572).[7]

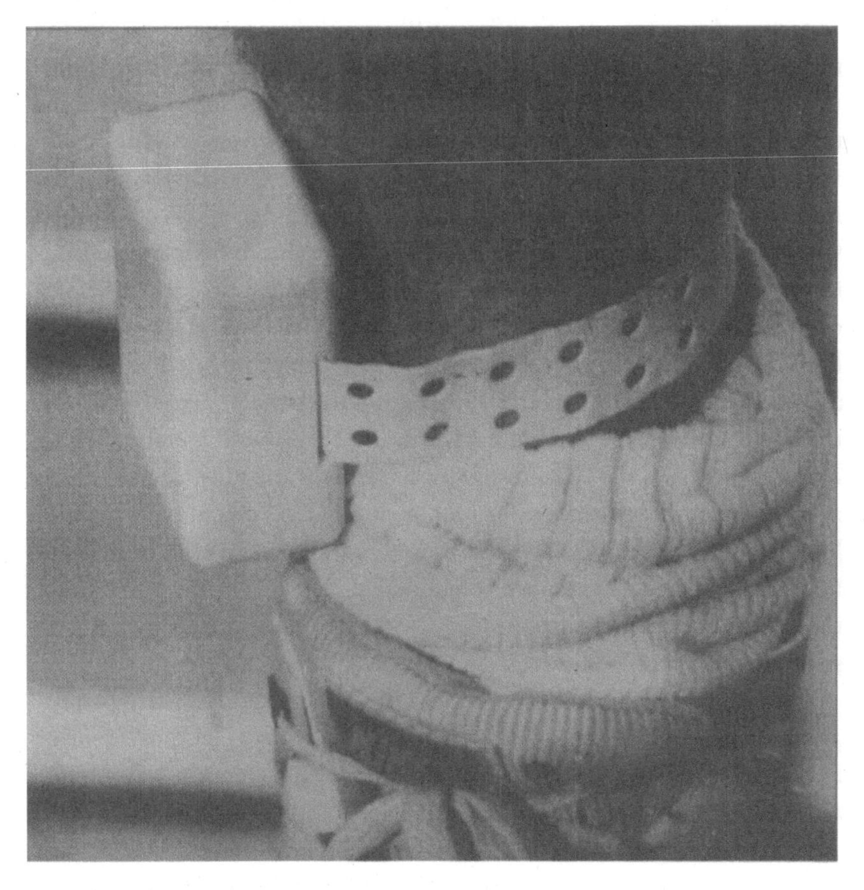

Figure 5.5
Ankle monitor used in 1990s in Sweden. *I DAG*, August 17, 1994.

Two current developments are the introduction of GPS-based monitoring instead of radio frequency and the discussion of extending ankle monitoring to minors instead of placement in institutions for juvenile offenders. The technical shift toward GPS-based monitoring still follows the earlier principle that the control room is alerted in case the offender leaves the confines of the home or moves outside of a specific area. There are two versions of the receiver: one that is installed at the offender's home and one that is contemporary smartphone and that should be carried close to the body at all times. The receiver and the sender that is still worn around the ankle are connected via Bluetooth. According to our informant at the Swedish Prison and Probation Service responsible for ankle monitors in one specific unit in Stockholm, Bluetooth-based units are much more precise, but there might be technical issues if a lot of different Bluetooth-based devices are used in the same area or household. In contrast to the radio frequency–based units, the Bluetooth signal cannot be enhanced, and there might be issues for the units to connect even within the home. The officers from the Prison and Probation Service can also send short messages to the receiver unit that includes a display where offenders can also check and follow up on their schedules. In the future, the plan is to also use the already-existing possibility of remotely requesting an alcohol check for which the identity of the offender is verified through facial recognition. This would save time for both the offenders who do not have to visit a specific office to be checked and the Prison and Probation Service taking the sample. Technologically, this is already possible and employed in the US. As with merely tracking offenders via GPS, the facial recognition feature is still missing a legal basis in Sweden. At the same time, the monitoring systems are prone to technological failures. In early 2021, ankle monitors at several Swedish open facilities malfunctioned and stopped working completely. The broad failure of the system was attributed to the COVID-19 pandemic that led to delivery shortages of devices and parts from China.[8]

Besides monitoring the offender at home, the system can also be programmed to exclude the offender from certain areas. The movement of the offender him or herself is not tracked. In 2008, a motion was submitted to the government from the parliament in preparation of GPS-based ankle monitoring.[9] It took until 2016 when the Prison and Probation Service started pilot projects for use of the GPS-based tracking system.[10] In connection with the introduction of GPS-based tracking, the Prison and

Probation Service also changed the commercial provider of the technology from Attendi to SuperCom, which has been used in open low-security prisons before. Now offenders can keep the same device while moving from low-security prisons back home to serve the remaining time of their sentence. GPS-based tracking requires legal adjustments that were under way in 2021. The proposed law is being discussed to allow for broader use of ankle monitoring in, for example, cases where the offender has not applied for ankle monitoring or was sentenced to prison.[11] The law was passed in 2022 and gives the Prison and Probation Service more extensive freedom to use ankle monitors. This change comes in a time when the Prison and Probation Service is under constant pressure due to overcrowding and is planning the largest extension of its facilities in Swedish history. Still, the Prison and Probation Service assumes that the building boom will not cover the expected needs and consider ankle monitoring as an efficient way of punishing more offenders.

The GPS tracking system encompasses not only a technological shift. The change coincides with a renewed discussion of punishment for juvenile offenders. Usually, underaged offenders with a high risk for recidivism are placed in so-called *särskilda ungdomshem* (community homes for juvenile offenders), where young adults with social and mental health issues are also placed. However, rehabilitation success has been slim. Underage offenders should now be monitored electronically while they stay at home, continue their educational programs, and remain within their family. At the same time, GPS-based monitoring would ban them from specific areas such as those with high gang-related activities so as to cut their ties with criminal networks. According to the probation officer we spoke with, the programming of the device can be compared to drawing area maps in Google maps. Furthermore, the new law would allow the use of ankle monitors for periods longer than six months, which is currently the maximum length. Additionally, ankle monitors could be used under probation and directly after offenders have served their prison sentence. Our informant jokingly remarked that with these three groups for new ankle monitoring, "the Swedish prisons will become a very calm place."

FROM SERVING TIME AT HOME TO THE BETTER SELF

With the increasing number of offenders who are serving their sentence at home, the prison is moving out of the confines of the prison walls.

Incarceration is mobilized and decentralized, but this is only one way of seeing the trickle-down effect of punishment from the prison into the home. Another aspect lies in the technological development where we can draw comparisons between ankle monitors and modern wearable technologies such as Fitbit and the Apple Watch (Bernard, 2019). While Andreas Bernard draws genealogical lines between ankle monitors and modern self-quantification and self-surveillance, we can also draw very specific lines between devices initially developed in and for the corrections sector that are migrating to other parts of society. There are different ways of tracking technological development. One way is to track the immediate sources of inspiration for certain devices or technologies through forward and backward citation in patents.

Like academic articles, patents are citing and being cited, and these citations can give us some indications about the history of the development of certain technologies. Surely some patents are merely submitted for reasons of prestige and internal tracking, but in general they have the aim to secure commercial outcomes of specific inventions in the future. Patents materialize intellectual property and constitute documents of openness and closure of innovation (Hemmungs Wirtén, 2019). Patents furthermore constitute media that capture and construct specific discourses and representations that are expressions of the contemporary mode of technological innovation (Shapiro, 2020) and hint at how specific companies are envisioning the future (Delfanti, 2019). But patents also reveal trajectories of technological development in particular through the citation practices. Depending on the specific jurisdiction within which the patent is submitted, citations are a requirement or not. In the US patents must cite immediate predecessors, while European Union regulations do not require citations. Hence, the number of citations varies depending on where they have been filed. In any case, the patent system constitutes a technological classification system that allows for the placement of inventions in connection to their inventors and companies in time and space. Through the citations it is possible to draw networks of affinity between different inventions that gives an indication of how technological developments are moving between sectors and industries. Zvi Griliches, for example, argued in 1979 that the frequency with which patents cite patents from different industries is an indicator of the relative proximity and distance between industries. Furthermore, patent citations tell us something about the centrality of certain inventions. As Jaffe and Rassenfosse

(2019) argue, "a patent with many or technologically diverse forward citations corresponds to an invention that was followed by many or technologically diverse descendants" (p. 1361). The Google patent database offers the possibility to track citations both backward (citations of predecessors) and forward (the patent in question being cited by other patents).

Taking the original patent of Schwitzelgebel for his first so-called behavioral supervision system with wrist-carried transceiver as a starting point, it becomes apparent that the idea of monitoring people through body-worn devices quickly migrated into other areas (figure 5.6). The patent number US3478344A from 1965 has in total 160 forward citations. Most patents—around 70—that cite Schwitzelgebel's patent are health-related patents (for tracking objects, patients, or beds in the hospital environment). The second-largest entity, with 30 citations, is a group with security-related patents including child tracking and alarm systems. The third most important group of patents are for technologies used in the work context, namely staff tracking, time, and accounting systems, with in total 19 citations. There is also a larger group of patents for telecom-related technologies, such as tracking the location of callers. This group includes such companies as the Swedish telecom company LM Ericsson (for acknowledging calls in a system for wireless staff locators) and Motorola (for a wristband). A group of 11 patents for criminal justice–related technologies including additional versions of remote monitoring of offenders is also included but does not constitute most forward citations for Schwitzelgebel's patent. A smaller group of patents includes those for tracking in warehouses and animal tracking (in total 11 citations).

Tracking citations of commercial applications for wearable technologies also reveals interesting links with ankle monitoring devices within the criminal justice system. One of the latest Philips patents for a biometric tracker, for example, includes both patent citations of commercial wearable products such as the Apple Watch and monitoring devices in the corrections sector. Based on the biometric tracker patent, Philips has launched a sensor for clinical surveillance for COVID-19.[12]

PRISONS AS TEST BED ENVIRONMENTS

Following the citations of monitoring devices for the criminal justice sector suggests that prisons can function as test bed environments for innovative technologies. This is expressed not only in the movement of patents

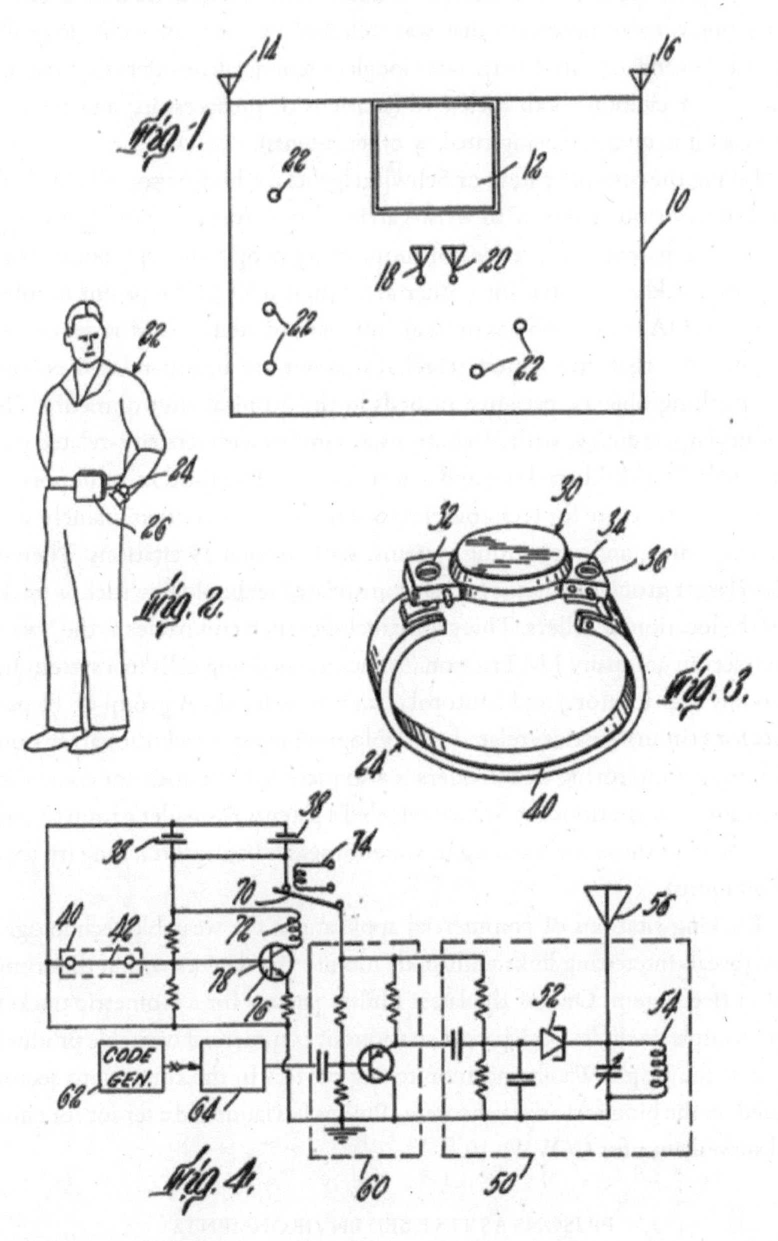

Figure 5.6
Accompanying image from the original patent for the behavioral supervision system with wrist-carried transceiver. US3478344A application filed 1965, patent published 1969. Google Patents (n.d.), *Behavioral supervision system with wrist carried transceiver*, https://patents.google.com/patent/US3478344?oq=behavioral+supervision+system+with+wrist+carried+transceiver

between different sectors but also in the plethora of best practice cases that are presented at the expos. Phillip Lowery, the national director of state government at Johnson Controls—one of the largest actors in the prison tech industry—opened the catalog of the 2019 edition of the American Jail Association expo with the statement that the future goal is to build "better jails through smart technology and training." In the best practice example, he emphasized smart technology to solve emerging challenges; "crumbling infrastructure, deferred maintenance, and rising utility bills are taking their toll on correctional budgets at the local, State, and Federal levels. And due to an ever-increasing population, spending on jails and prisons continues to demand a larger share of tax dollars. With increasing needs and shrinking budgets, correctional leaders must look for smarter ways to improve facilities and ensure safety for offenders, officers, and the greater community."[13] The prime solution to all these emerging challenges follows swiftly: "one solution . . . is the creation of 'smart' jails and prisons. A smart correction facility creates a performance infrastructure that embraces emerging technologies to enable safer environments for correction officers and inmates."

The technological solution to the crisis of the American criminal justice system is mirrored in other sales pitches by prison tech companies as well. UniLink, sporting the slogan "Excellence," argues that "technology in prisons has the possibility to transform the prison service and outcomes for the incarcerated as well as to drive efficiencies across the whole service and to the society on a much broader level" in one of its brochures that was distributed at the American Jail Association expo in Kentucky. At the same expo, one of the largest and most impressive booths promoting Offender360 by DXC Technology aimed to "reform through performance" by offering software solutions for jail and prison administration including record and assessment management. The software offers solutions to implement data-driven decision making in the prison facilities using real-time visualizations and analysis. With the help of systematizing data collection and analytics, the software promises to automate the detection of anomalies and trends in prison population behavior, which allows staff to spend more time acting on issues and problems within the detention facilities. To illustrate the potential of data analytics in prisons, a salesperson pitched the visitors a story of a facility that implemented data analytics by including a systematic analysis of reports on incidents by locality. Very quickly the software was able to detect a pattern. Many incidents were happening in locations with televisions. Consequently, the administration removed all TV sets, and the incidents went down almost immediately. The software was at the time of the expo only used by around 20 facilities in the US. The fancy outlook of the booth, including some of the expo's most popular merchandising and giveaway bags, implies the aim to grow quickly, and the delegates were impressed. One of the delegates from a small US facility that joined one of the test shows said happily, "So now I can finally fire my coder?"

The imaginary of radically transforming prisons with the help of smart technology is translated into initiatives to implement tech test beds for radical innovation. During the expo and conference Digital Transformations in Lisbon, the chief information officer of the Swedish Prison and Probation Service referred to the KRIM:Tech initiative, illustrating the test bed and radical innovation imaginaries connected with prison tech. The technology hub describes its aims in the following manner: "KRIM:Tech is long term. We test and we change. We work in teams and help each other, smart technology, and the latest research. We get started and restart, develop and evaluate. KRIM:Tech is a big change for public authorities in Sweden. And the only way to contribute, the only way to be part of this change is that it becomes part of us."[14]

The technology hub KRIM:Tech of the Swedish Prison and Probation Service was set up to radically innovate the organization primarily with the help of smart digital technology. Between 2018 and 2019 the hub received funding from the Swedish innovation agency Vinnova to pilot test bed activities for digital and radical innovation.[15] Besides this very explicit project to develop test beds within prisons, there are as we show numerous previous examples of test bed work, or what Fred Turner (2016) has called "real-world prototyping" in prison facilities.

At the same time, the development, prototyping, and test bed work in prisons can be linked to broader discussions of innovation and development that often involve discourses of disruption, creative destruction, and failure as a pathway to success, as Appadurai and Alexander (2020) argue. The innovative nature of digital culture has traditionally been connected with the liberal spirit of what Turner (2006) has called the "new communalists" in Silicon Valley, emphasizing changing the consciousness of individuals rather than political systems. This depoliticized counterculture provided the futile ground for the new economy driven by and driving digital innovation. While the spirit of Silicon Valley quickly incorporated both failure and the imperative to radically innovate in its ideology, innovation and technological development are always also dependent on mundane tasks including testing, repairing, and maintaining, as scholars of science and technology studies have emphasized for a long time (Jackson, 2014; Roberts, 2019; Sawhney & Lee, 2005; Velkova & Kaun, 2019). Mary Gray and Siddharth Suri (2019) explore, for example, the hidden ghost work that is essential to the development of artificial intelligence, including simple categorization tasks and content moderation. These tasks are often outsourced and take the shape of microtasks facilitated by platforms such as Amazon's Mechanical Turk. In this context, science and technology studies have suggested shifting the focus from pioneer communities (Hepp, 2016) and centers of innovation to the underbelly and the shadows of digital culture and digital economies as well as focusing on the contributions of mundane acts of expected and unexpected use that feed into innovation and development work. Test beds and living laboratories accordingly involve both invisible and visible forms of work by those steering and implementing experiments, testing, and test subjects (Star & Strauss, 1999). We are here particularly interested in contributions that often go unnoticed and unseen and that encompass limited degrees of freedom for the laboring subject. In the context of labor, science and technology studies emphasize practices of maintenance and repair to counter the focus on creative work and innovation that often dominates design discourses. Instead, research within the field has highlighted the invisible work of low-wage workforces with limited autonomy and power and their contributions to technology and science through, for example, maintenance and repair, often rooted in feminist ethics of care (Denis et al., 2015). Accordingly, several studies have explored the role of maintenance and repair for different kinds of infrastructure, technology, and science; for example, exploring work that is needed to keep a road, building, machine etc. in good condition and advocating a shift in perspective from innovation to brokenness and how to deal with it (Jackson, 2014; Mitrea, 2015). This perspective on maintenance work of the technological underclass can be extended with the contributions of mundane users to technological development. Appadurai and Alexander (2020), drawing on Ivan Illich's notion of shadow work, argue that

> we are no longer just "users": we have become testers, analysts, and feedback providers at the very heart of real-time design improvement. Thus, despite the fact that endless apps will quickly become obsolete, there is no such thing as an

> absolute "failure" in the world of mobile apps. Instead, the app economy relies on an endless communication loop between end user and product designer, so that each failure is simply the driver of the next improvement. (Appadurai & Alexander, 2020, p. 55).

In that sense, we move from an emphasis of radical innovation and disruption of and by technologies toward an approach that highlights the distributed and socially embedded character of technology, toward environments where slow, work-intense development takes place (Ruppert et al., 2013).

Appadurai and Alexander (2020) dissect the unpaid free labor of all users who contribute to the development, repair, and maintenance of digital media. Our argument relates instead to more specific design approaches to innovation and testing, namely test beds and real-life laboratories, also called living labs, that advocate user-centered and open-innovation ecosystems for design development (Engels & Rogge, 2018). The history of real-world labs is often traced to MIT's living lab—the so-called PlaceLab—that between 2004 and 2007 operated a residential living laboratory in a multifamily apartment building in Cambridge, Massachusetts. The building was equipped with hundreds of sensors to semiautomatically document the activities and movements of residences.[16] The approach of living labs to innovate and develop information and communication technologies was quickly taken up by different design schools and was further developed into different substrands, including long-term user involvement (Ogonowski et al., 2013) often involving public-private partnerships (Niitamo et al., 2006). The approach of living labs is currently further developed in the smart city context, including the now infamous Sidewalk Lab by Google's Alphabet that was running the much-discussed and -criticized pilots in the city of Toronto.[17] In contrast, test beds are often considered to be less participatory. Rather than integrating the user in participatory and coproductive ways, test beds focus on the development of isolated features of hardware and software. The Swedish governmental innovation agency Vinnova defines test beds as a physical or virtual environment within which companies, academic institutions, and other organizations collaborate to develop, test, and implement new products, services, processes, and organizational solutions for specific areas.[18] Vinnova distinguishes between laboratory environments, simulated environments, and real environments. Universities and research institutions as well as industry are mainly working with the first type of test beds, while the third type involves the public sector to a larger degree.

While Appadurai and Alexander's (2020) argument that we—as users of digital technologies—all become shadow design workers who labor for free toward innovation, others have argued that scientific and technological innovation is often tested among particularly vulnerable populations: the sick, the poor, and the marginalized (Preciado, 2013; Rusert, 2019). In that sense, different forms of testing in prisons, including medical trials, can be considered earlier versions of such real-life or living laboratories that link to a much longer history of conducting research and experiments among vulnerable populations, that "the procedures adopted for the captive flesh demarcate a total objectification, as the entire captive community becomes a living laboratory" (Spillers, 1987, p. 68) referring to several ways in which enslaved communities were exploited for, among other things, medical research. Britt Rusert takes this as a starting point to engage with the Tuskegee experiments that focused on syphilis infections. Part of the experiment was to hold back treatment for certain test groups with infections who were recruited mainly among the African American population. Rusert argues that "Tuskegee test subjects quite literally served as natural resources mined in the government's quest for biomedical dominance on the global stage, connecting an apparently insular and exceptional experiment in the Deep South with the transnational commercial networks of US science" (Rusert, 2019, p. 37). She argues further that experimenting with humans has long been mainly dependent on underdevelopment. Similarly, Helen Tilly develops a history of scientific testing in Africa during colonial times. The continent as a whole was considered a living lab not merely metaphorically but also literally, agreeing with the idea that manipulation, control, and experimenting were not only acceptable but explicitly desired. She argues that "Africans and their environments were in this sense captive subjects of such experiments" (Tilley, 2011, p. 11). Development and innovation in the global North have depended on the underdevelopment of Africa, where potential test subjects are located, but also marginalized populations within the global North.

INCARCERATED INDIVIDUALS IMAGINING PRISON TECH

Companies' view of prisons and the corrections sector as valuable avenues is one side of the entanglement between technologies and punishment. The perspective of the incarcerated themselves is another. Although rarely acknowledged, incarcerated individuals are also part of audiences for socio-technical imaginaries as well as popular media representations. As Knight and Bennett (2020) have shown, imaginaries of communication technologies and reception of media content should be considered in terms of agency and sometimes even resistance. One way of expressing their imaginaries of technological development in the prison complex is prison papers. As we have seen in relation to prison media work, prison papers are a forum for developing and expressing the perspective of incarcerated themselves. The content of the prison papers naturally changed over time during the twentieth century. Sometimes the changes were rapid and haphazard, not least since the members of the editorial boards constituted by the incarcerated were often changing as a consequence of releases or transfers to other prisons. However, the content also mirrors general social and cultural developments as well as the changes in penal ideology that took place over the course of latter half of the twentieth century. While the contributions of prison papers often focused on the immediate everyday lives of the incarcerated including the quality of food and access to meaningful activities, there are also contributions that directly engage with prison technologies. Figure 5.7 is a drawing from 1962 depicting the automation of the court room. Instead of the judge, a machine is imagined handling the sentencing.

The text below the drawing reads as follows:

> The computer programmer is the judge in the middle of the picture. He has just fed the machine with the offender's minor felonies (the little oppressed man on the chair). At the table sits the prosecutor, the ragtag representative of order in society. It is on the call of this man that much of the data is based with which the computer machine is fed. There are feeding holes for the pros and cons of human beings, for the warm and human views of the defense as well as for the claims of the prosecution. The machine sifts out irrelevant data, the judge just feeds and feeds—and out comes a clear and printed verdict, which no one can appeal. The verdict is pronounced by the leftmost arm of the machine. (*Hallbladet*, 1972, No. 2)

In this image, the judge becomes merely a mediator who feeds information into the computer, while the computer delivers a decision without the possibility to appeal. The depiction of the automation process made it into the prison paper *Hallbladet* at a time when the Swedish public agencies and institutions were preparing to introduced computers on a large scale. The Swedish Prison and Probation Service, for example, started to

Figure 5.7
Illustration of how incarcerated have imagined computer-based automation of the criminal justice system. Published in *Hallbladet*, 1972, No. 2.

digitize its work in the beginning of the 1970s, although the public discussion of automation was quite intense already in the early 1960s (Rahm & Kaun, 2022), and the 1970s evolved as the peak of digitalization of public administration, forming what Johan Fredrikzon (2021) has called the era of early digitalization. Human decision making becomes a minor part of the verdict for the offender. Accountability and responsibility are delegated to the machine instead. The automation of the courtroom extends to other discussions of alienation within corrections that incarcerated individuals return to time and time again in contributions to prison papers. The loss of autonomy is here related to the automation of decision making with the help of machines, while other contributions discuss the loss of temporal autonomy, the ability to decide about one's own schedule and structure of daily routines. However, the loss of autonomy also highlights an alternative sociotechnical imaginary that is not present in the discourses produced by prison and tech company officials, namely what happens with human agency in relation to technology, what the dark sides of automation are, and what happens to offenders and incarcerated individuals in the process. These imaginaries emerge at the margins of the broader discourses of technology implementation and development in the prison context, but they challenge the hegemonic ideas of the good of technology in crucial ways.

PRISON MEDIA TECHNOLOGY: CARCERAL TEST BEDDING

The tropes of sociotechnical imaginaries of prison tech discussed here illustrate the ways in which prisons emerge as milieus for punishment but also technological development. Rather than specific work practices of guards and incarcerated persons or their experiences with prison tech including forms of resistance, we focus on how prisons serve as test beds for technology development. This includes how specific technologies are developed for and in the prison context and are then dispersed into other areas, such as the health and wellness sector. We highlight what kind of sociotechnical imaginaries are developed by technology mediators of prison tech who gather at expos and fairs, showing how technological advancement is taking shape and entangling material and discursive aspects. These sociotechnical imaginaries of prison tech that are developed, presented, and negotiated by the technological mediators at the expos are performative, as they are translated into concrete projects including the implementation of test beds for radical innovation.

Test bedding in prisons encompasses the production of a sociotechnical imaginary of backwardness that is a precondition for radical innovation of not only parts of specific facilities but also the institutions for punishment. Prisons thus provide one context for technological advancement next to a plethora of other high-stakes environments such as the welfare, health care, and education sectors. High-stakes environments concern especially vulnerable populations that are strongly affected by technological and organizational changes. At the same time, these specific environments are the foremost places for testing technological fixes, such as social benefit provisions and child welfare that Virginia Eubanks (2017) has argued for. Technologies that are tested in prisons are irrevocably entangled with vulnerabilities produced in and by incarceration. Highlighting the entanglement of the "pain of imprisonment" with technological development nuances positions that often one-sidedly underline the empowering features of technology. Making the experiences of incarcerated people visible questions the assumption that all technological development is inherently moving us toward better societies. Instead, we should ask where technological fixes are producing new and reinforcing vulnerabilities rather than resolving them. This means fundamentally questioning the idea that technology is always the best way forward.

6

CONCLUSION: PRISON MEDIA AND MOBILE INCARCERATION

The novel *Grundbulten* (The foundation bolt) ends with a prison riot scene. During the riot a group of incarcerated persons take guards hostage, which is immediately picked up by the news media. The rioters follow the media reporting closely throughout the occupation. One of the incarcerated individuals is elected as the designated press spokesperson, and a "transistor radio in a corner updates [them] continuously about the development of the view of the outside world" (Ahl, 1974, p. 242). The protesters, however, not only strategize about the next steps but also finish a poker game they had started just before the riot broke out. This mixture of a total state of exception—the riot—and the everyday of prison life—the poker game—is not only the climax of the book but also the metaphor for prison itself, and so it is for prison media. The prison is far removed from society, a heterotopian state of exception while also filled with routines, a constant stream of everyday activities that moves on regardless. And along these two poles constituting the prison experience there are media or, more specifically, prison media: media that are particularly developed in and for the prison and the prison as a medium. Prison media constrain and enable communication but also constitute a bridge into the lives beyond the prison walls, as the reporting about the riot that connects the incarcerated with the world outside the facility illustrates.

Prison media are materials, devices, and technologies but also imaginaries, representations, and symbols as well as practices. And in many ways, prison media are entangled with fundamental questions of sociality, how people are brought together and how they are separated by communication and how distinctions and boundaries but also commonalities and cohesion emerge. Prison media serve as an example and a focal point to consider how social worlds are connected and disconnected from each other. Ultimately, we argue that prison media are media technologies proper but also a heuristic, a way of seeing and making social processes in relation to media technologies visible. The story of prison media emerges in a contested field that brings together histories of media technologies with histories of institutions of normalization, control, and power such as the prison. Media are part of the normalization strategies that have been ascribed to prisons, hospitals, and schools, following Michel Foucault, but at the same time media potentially question forms of incarceration and contribute to changes in disciplining and punishing. The Swedish industrial prisons of the 1960s and 1970s were mainly driven by the idea of rehabilitation through work. The production of hardware for media infrastructure consequently became a natural part of the idea of normalization through work routines, as we discussed in chapter 3. Similarly, media technologies are produced to safeguard prison facilities and realize the ideas of punishment. Here prison technologies are a prime example of media produced for the prison context that quickly find areas of application outside the corrections sector, as we showed in chapter 5. However, prisons emerge also as media themselves through their architectural features and ways of structuring communication, as we discussed in chapter 4.

Prison media emerge around the poles of visibility and invisibility. Prison media as buildings represent changing penal regimes in a visible manner. We have touched on the first modern cell prisons that were often placed in the center of the town, reminding of the power of the penal regime. The Swedish industrial prisons of the 1960s, though not situated centrally, expressed the crucial idea of normalization through structured work in their architectural form. Similarly, the move toward decentralized smart prisons and mobile incarceration with the help of ankle monitoring has produced a visible expression of shifting penal regimes. At the same time, there are many aspects of prison media that are largely invisible. The work that produces prison media and the practices of testing surveillance technologies in prisons are hidden from the public eye. However, it is not only a tension between visibility and invisibility that is characteristic of prison media. There is also a tension between the "inside" and the "outside" at which prison media are situated. Prison media are foundational to constituting and making possible forms of punishment and rehabilitation; in that sense they are integral to penal regimes and constitute the inside as something secluded from the social world. At the same time, prison media connect the inside with the outside, not only through bringing news and messages to the incarcerated individuals from the outside; indeed, prisons have become hyperpresent in media culture, constituting what Michelle Brown (2009), among others, has called the "prison spectacle."

Taken together, prison media should be considered in dialectical terms. They emerge as places of rehabilitation and punishment, states of exception and everyday life, and are removed from other social worlds while being also an integral part of society. This ambiguous character of prison media leads us also to the fundamental question of how prison worlds and other social worlds are linked or disconnected. Erving Goffman (1961) has famously explored the ways in which patients and incarcerated individuals—by entering the "total institution"—are trained in their social role as patients and incarcerated: they are institutionalized. Part of the institutionalization is the loss of individuality and subjectivity. Incarcerated individuals are assigned numbers instead of names, and their personal history is transformed into a treatment scheme. The total institution hence establishes a world that follows specific social roles that need to be learned, trained, and practiced. In the establishment of the prison world, media and communication are important parts. Access to communication and media is highly regulated and hence differs fundamentally from the ways in which people connect with and through media in other social worlds (Jewkes & Reisdorf, 2016). Similarly, there is currently a discourse on the technological backwardness of penal institutions to promote smart prison initiatives (Kaun & Stiernstedt, 2020a, 2022). Here, the prison world is characterized as lagging behind the general development of digitalization and therefore constitutes a fundamental outside of other social worlds. Access to digital media in prisons has been framed as a human rights issue and points to the crucial consequences of depriving incarcerated individuals from access to communication (Scharff Smith, 2012b). As Victoria Knight (2016) has shown in discussions of introducing certain media technologies to the prison context, the penal system and its relations to discipline and punishment crystalize in debates around media access. However, rather than being about media technologies as such, the often heated and controversial debates point to the ways in which societies negotiate social norms and boundaries between "the good and the bad," the normal and the deviant. These perspectives establish prisons as a separate social world that follows fundamentally different logics than other social worlds. The emphasis is put on differences and disjuncture that separate prison worlds from other worlds.

In contrast, ethnographic inquiries such as Didier Fassin's major work *Prison Worlds*, scrutinizing the French penal system, emphasize how logics of incarceration are entangled with economic and social inequalities in society. Prisons and the penal system manifest and reinforce social divides, Fassin argues. While taking the excruciating experiences of incarcerated seriously—the pain of imprisonment, experiences of isolation, emptiness, meaninglessness, and lost time—he emphasizes that prison worlds crucially reflect and extend forms of racial and social exclusion prevalent in French society. According to Fassin (2017), the prison is only one expression of the punitive state that "extend[s] within and around the prison, through both the carceral and the correctional logic" (p. 287). In parallel to that, the prison has been mobilized in metaphorical or discursive terms. Following Foucault's understanding, the modern prison emerges in the context of new ways of scientific knowledge production that encompasses the distinction between normal and deviant. The deviant—the criminals and mentally ill—need to be treated, normalized, and confined. Hence, there are modern rationales that motivate incarceration. Similarly, Stanley Cohen and Laurie Taylor (1976) take their starting point in the prison context for exploring practices and theories of resistance to everyday life. As a follow-up book to their seminal study of the effects of long-term incarceration in maximum security prisons (Cohen & Taylor, 1972), they argue that the consciousness of incarcerated is occupied with fundamentally different matters than in the outside world. Matters such as passing of time, relation to work, privacy, and deterioration are experienced both inside and outside the prison, but they are attended to quite differently. They have a very different salience and urgency for the incarcerated. In that sense, Cohen and Taylor argue that prisons are magnifying glasses and intensifiers for matters that pass invisibly in the outside world but still matter for everyday life. Following Foucault's approach of exploring the prison as a model for sociality more generally, several popular science publications invoke the prison as metaphor to consider emerging forms of surveillance and control with and through digital media. The German journalist Adrian Lobe (2019) engages with the data prison built and maintained by users on a voluntary basis. The journalist and media theorist Andreas Bernard (2019), similarly, invokes the prison to develop an understanding of contemporary digital culture. Bernard traces the genealogy of social media profiles back in history to its origins in criminology and practices of profiling of deviant individuals. He also traces the history of ankle-worn trackers and connects them to the idea of monitoring incarcerated individuals with the help of ankle bracelets. Rather than tracing the specific technological developments, he emphasizes discursive trades and ideas of surveillance that are remediated in popular digital devices in nowadays. Bernard hence establishes contemporary digital technologies as the grandchildren of criminology and psychiatry and shows the ways in which ideas of control and punishment are dissolved into the social world as a whole through these digital devices.

What we can conclude is that there is a tension between two approaches to prison and punishment, between Goffman's total institutions and Foucault's self-inflicted prison and total surveillance, as Didier Fassin (2017) argues. While Goffman establishes the prison as a place that lies outside of society and follows its own rules and habits, Foucault considers the prison as a form of punishment and surveillance that has bearing for society as a whole. According to Fassin, we need to invert both perspectives, namely questioning the total institution as being total and acknowledging the multiple ways in which the outside worlds take themselves behind the prison walls through the law, private enterprises, television programs, contraband, and so on. We also need to question that the prison dissolves completely into the social space. There are very specific practices, routines, and lived experiences of the incarcerated that emerge in the prison, such as prison slang and specific social hierarchies based on the offenses and prison sentences. Even if the metaphor of the prison to analyze and describe more general practices and discourses of surveillance is convincing and there are obviously historical overlaps and genealogies, the experience of incarceration, the pain of imprisonment and deprivation, remains a very specific one. Acknowledging both, Fassin suggests considering the carceral condition dialectically, "both as it is traversed by the reality outside, and as it remains an irreducible fact" (2017, p. 295). He concludes that "the carceral condition is not the world's shadow because of those who are locked up in prisons . . . but [instead] because of the social inequalities it contributes to reproducing and the everyday injustices it helps to legitimize."

Similarly, histories of media technologies and prisons are entangled in multiple ways while also being separate from each other. Over time they evolve in parallel, moving toward each other and then diverge again. These connections and divergences are not arbitrary. Instead, they tell us something about the ways in which societies change and where the social fabric is porous, dissolving, and delicate by pointing to an ongoing negotiation as to whether prisons are a shadow of social norms (Fassin, 2017), unite all aspects of the social including ideas of the desirable and undesirable as a heterotopia (Foucault, 1984), or are far removed, isolated, extreme places of their own social rules (Goffman, 1961).

In the context of prison media work, this means that many forms of work are characterized in one way or another by unequal power relations within (and outside) the capitalist labor relations of modern society. These unequal relations become especially apparent in the prison context. While prison work in the Swedish case became increasingly central as a tool for rehabilitation and normalization of prison life, often the working conditions were lacking creative and intellectual challenges and were experienced as dull. Considering prison media work makes the entanglement of state investments in media infrastructure development visible, as it provides free and cheap labor in the prison facilities for corporate as well as state-owned companies producing and developing media. Prison media work hence adds another layer of invisible work "below the line" of the creative industries. In a similar way, prison architecture encompasses these connections and disconnections between life behind bars and beyond. As architecture more generally (Colomina, 1994), prison architecture constitutes a medium that allows for, channels, and constraints communication in very specific ways. This mediating role of architecture is extended by the entanglement with communication technologies that allows certain architectural features to emerge. For example, the culvert system in Swedish high-tech prison facilities primarily built in the 1960s and 1970s was made possible through CCTV technology. Video surveillance and electronic lock systems allowed for the automation of moving incarcerated individuals around in the large-scale prison facilities. The early use of these technologies at prison facilities allowed for further refinement and adjustment before they were used in other contexts as well. This form of test bed technology in prisons is another important way of entangling carceral logics with social logics beyond the prison walls. Often presenting the prison as lagging technologically behind general societal development, public-private partnerships draw on prison facilities as live laboratories that are rarely constrained by privacy concerns.

In connection with these observations, we suggest that prisons are an integral part of social life while being removed and outside what could be considered most people's everyday life while being entangled with and part of the social fabric. And so are media. Media are both constituting our everyday lives and providing disruptive worlds beyond our immediate surroundings. Media have been increasingly conceptualized as deeply entangled with our lives, and it has even been argued that we are living media lives rather than living with media (Deuze, 2012). At the same time, media and media technologies are still connected with a fascination and air of secludedness; they constitute their own worlds that many aspire to become part of, ranging from celebrity to start-up culture. We suggest that we must think those intricate links and disconnections together in order to understand both prison and media worlds. In this line of thought, we follow calls by scholars of science and technology studies including John Law and Annemarie Mol, who suggest considering multiplicities of the social instead of searching for the one discourse that unites everything and forms the social world. Law (2013) argues that "first, we should treat it as a set of patterns that might be imputed to the networks of the social; second, we should look for discourses in the plural, not discourse in the singular; third, we should treat discourses as ordering attempts, not orders; fourth we should explore how they are performed and told in different materials; and fifth, we should consider the ways in which they interact, change, or indeed face extinction" (p. 95).

Based on Law's argument, Mol explores the associations and different modes of ordering. Mol speaks of forms of coordination between objects and subjects, between networks of the social. These connections or—in Mol's terms—"forms of coordination" are partial. They "allude to what, not in itself but through the act of comparison, appears to be both similar and different. Not like a single large cloth that is cut into smaller pieces after which the lost unity is simply a form to be sought. Not a functional unit nor an antagonistic opposition. But an inside and outside" (Mol, 2002, p. 80). In our context this means forms of coordination between the development of the penal system and the media technological system. These forms of coordination and multiple connections between the two systems are not arbitrary; they are deeply political and marked by power. As we show, similarities between prison worlds and other social worlds are invoked in very specific contexts and historical configurations, while their disconnects similarly are emphasized under very specific conditions. We consider this entanglement of media and prison histories in Marilyn Strathern's (1991, p. 35) terms as "more than one and less than many."

FUTURE VISIONS 1: THE SMART AND FULLY AUTOMATED PRISON

The focus of this book has been on historical accounts of prison media, but where are prison media headed? What is the future of prison media? Prison media represent not only different aspects of the social fabric but also different temporalities, including ideas and imaginaries of future developments that crystallize around the notion of the smart and fully automated prison. The idea of the smart prison needs, however, to be situated in a broader context of multiple temporalities of prison media. Prison media emerge in the context of tensions between the permanence of architectural infrastructures—prison buildings are built with a life span of several decades if not centuries—and specific inbuilt infrastructures that are bound to last for much shorter periods especially digital infrastructure, which is increasingly considered in terms of its impermanence (Brodie & Velkova, 2021). Digital innovation is often connected with imaginaries and expectations of fast changes, radical disruptions, and fundamental change. Hence, prison media emerge as a site of tension between diverse, multiple temporalities.

One case in point is the discontinued project of producing AI training data sets in a Finnish prison that was strongly welcomed by the Prison and Probation Service of Finland. The project was abruptly cancelled when the start-up company Vainu, which was the industry partner, was acquired by another company. In the changed portfolio, there was no room for the collaboration with the Finnish Criminal Sanctions Agency where incarcerated individuals were supposed to categorize images for training data sets (Lehtiniemi & Ruckenstein, 2022). Technology test beds in prisons might emerge quickly and then disappear just as fast when the political discourses as well as investments from private partners change. While the digital industry moves fast and changes quickly, publicly funded institutions such as prisons are built slowly and with the aim and need of permanence.

In relation to and extending the discussion of multiple temporalities of prison media, the prison environment is characterized by an ambivalent relationship to digital technologies, which could be described as a paradox of a data-intensive environment observing and documenting the incarcerated people constantly and consequently producing large amounts of data combined with often low digital technology access for the incarcerated individuals themselves. Digital technologies are increasingly employed for surveillance purposes but are more controversially discussed in the context of rehabilitation and access by the incarcerated individuals (Jewkes & Reisdorf, 2016). In that sense, the prison reproduces and amplifies the divide between smartification on the one hand and exclusion from smart technologies on the other that have been observed in the smart city context (Kitchin, 2014). Prison media emerge in this ambivalent combination of permanence and impermanence of infrastructure, access, and exclusion. Prison media illustrate that infrastructures are both permanent and impermanent depending on context and their specific use. Exploring these diverging infrastructural temporalities might also require a different methodological approach. For us it has been the decentered way of researching prison media that made the different temporalities visible. Rather than focusing on one specific infrastructure, we have come to understand the prison as an assemblage of different infrastructures that are built on top but also against each other, an infrastructure of infrastructures.

The emergent temporalities of prison media have to be related to the experience of time and the loss of temporal autonomy for incarcerated individuals. Prison life consists of endless repetitions and the temporal ambivalence of the incarcerated individuals' experience of on the one hand suspended "normal" lives, while on the other hand their bodies and identities age (Medlicott, 1999). The structure of time in prisons differs considerably from that of everyday life and is characterized by a suspension of the life outside along with endless repetition and strict schedules imposed on incarcerated individuals (Meisenhelder, 1985). People who are incarcerated are losing their temporal autonomy as they are disciplined through these time-based routines and completely exposed to external authority regulating their time use that is translated into excessive counting and observing of incarcerated individuals. The experience of waiting in a prison visitors' center becomes, for example, part of "pains of imprisonment" that encompasses both visitors and incarcerated individuals. The painful temporality of the prison is extended beyond the walls into the lives of family members and friends of people who are incarcerated (Foster, 2019). The waiting and loss of temporal autonomy in particular emerges out of uncertainty and unpredictability, especially during custody and detention . In this context, Kotova (2019) develops the notion of temporal pains of imprisonment to analyze the experiences by female partners of long-term incarcerated individuals in the United Kingdom. She argues that prison time extends beyond visitation time and that research also needs to consider the deprivation of mundane family experiences because of the prison sentence.

The processes of adaptation and change over time are particularly important and are captured in temporal metaphors such as dead time, out of time, and doing time that represent the permanent state of waiting and suspension of life. The penal power of the prison is expressed exactly through the disciplinary function of waiting, while the temporality of waiting is translated spatially into the prison cell (Armstrong, 2018). Incarcerated individuals are, however, punished not only by being put into the specific space of the prison (and the prison cell) but also in terms of compartmentalized temporality; that is, time is sliced up into spatiotemporal boxes that line up into a linear narrative of punishment, prison sentence, and rehabilitation. However, the key point of the prison and other disciplinary institutions is not to produce the ideal subject as an existing one but instead to reproduce the idea of this subject that is impossible to be realized (Armstrong, 2018). Hence, this impossibility of achieving the ideal subject becomes the productive motor of repetitive and eternal processes of control. The future in the prison context is often only implicated in terms of rehabilitation. These layers of temporal experiences by incarcerated individuals are complicated with smart digital technologies that emphasize temporalities that run counter to the lived experience of time.

There are three major temporalities that are emphasized in the discussion of smart digital technologies used in the prison context: (1) velocity and speed that are linked to real-time data gathering and analysis, (2) prediction, and (3) preemption. First, real-timeness refers to the appeal to track practices, developments, environmental changes in high speed. Velocity refers to tracking rather than analysis, as it takes time to compare the recorded data against previous data sets (Cheney-Lippold, 2017). However, the velocity of real-timeness is not as straightforward as it seems (Weltevrede et al., 2014). Different platforms and applications produce different temporalities and rhythms of real-timeness of devices and digital platforms. Hence, the notion of real-timeness of datafication is much more complex. The dominant temporal regimes linked to smart digital technologies are the seemingly future-oriented temporalities of anticipation, prediction, and preemption that are a result of real-time tracking. Anticipation refers to the act of looking forward to a later action by relying on data of past behavior.

According to Ben Anderson (2010), anticipation is concretized through two "anticipatory practices": prediction and preemption. Prediction refers to foretelling the future based on observations and experience, which is perfected in the context of smart digital technologies, since the likelihood with which a certain development will appear can be calculated on ever larger amounts of data. Preemption refers to practices and acts that prevent certain developments and behaviors from taking place. The aim is to forestall and preclude harmful threats. Based on predictive calculation, preemptive actions are also supposed to prevent concrete determinants from emerging in the first place. Following these definitions, the anticipatory practice of prediction is a presupposition for preemptive activities. Calculations based on big data that are at the heart of anticipatory practices seem to be addressing the future, while prediction is inherently conservative and past-oriented, as it is based on historical data (Cheney-Lippold, 2017). Anticipatory practices such as prediction and preemption, rather than being future-oriented, hence reproduce and reinforce assessments and decisions made in the past and contribute to a programmed vision of the future.

Smart digital technologies are seemingly about future-oriented temporalities; however, anticipation, prediction, and preemption constitute a multiplication of the present rather than the future (Coleman, 2018). The temporalities of smart digital technologies are in that sense not about projecting a visionary future that needs to be actively constructed but instead reinforce established models and analyses. Rather than being concerned with the future, temporalities of smart digital technologies seem to be trapped in the past and present.

From the outset, smart media technologies create an increasing desynchronization of the different speeds of prison time. For the individual, the experience of doing time is often one of slowness, routine, waiting, and the general—and intentionally created—feel of time (one's individual lifetime) as inhibited or put on hold. The smart digital technologies increase, on the contrary, the pace of data collection and processing that is—and always has been—a part of how time is "done" to the incarcerated individuals. When procedures of gathering and making sense of data are increasingly automated and integrated in smart systems, they can furthermore produce immediate results and actions. For example, a deviant pattern of movement might lead to the automated response of a lockdown within a smart CCTV system. The fast pace of technological temporal regimes thus paradoxically serves the purpose of increasing or at least maintaining the slowness, routine, and inhibition of time that are the existential experience of incarcerated individuals, since the purpose of surveillance and control is to make sure that nothing (unexpected) happens. Unexpected means here that the prison institution is organized based on programmed future visions of correct behavior of both incarcerated individuals and guards that are enacted with the help of smart technologies. This programmed future vision of the prison relates to the dimension of smart surveillance and control and the increased importance of past actions for the patterning of incarcerated individuals' existence and for future predictions. Since many different kinds of data are not only collected but also stored and continuously analyzed and since smart systems rely on predictions from past patterns, this temporal dimension becomes increasingly important for the management of incarcerated individuals and the predictions of their future that is used in attempts to rehabilitate them and help them adjust to future life outside of the prison. Paradoxically, then, the speeding up and increased efficiency that are promised by smart technologies reinforce the experience of trapped temporality of incarcerated individuals while they remain objects rather than subjects of emerging temporalities of smart technologies and are included in "the digital imperative" only one-sidedly, namely through a focus on control, monitoring, and surveillance rather than rehabilitation. The future of prison media seems currently to be biased toward an imperative for surveillance and control and only rarely integrates future visions of prison media for rehabilitation.

FUTURE VISIONS 2: MOBILE INCARCERATION

The contemporary penal regime has, in this book as elsewhere, been described as an era of mass incarceration. Since the 1980s, the number of incarcerated individuals in many jurisdictions has been rising. This holds especially true for the US context, but mass incarceration is a worldwide trend. Prison sentences have become longer, and the prison is discursively depicted as a solution to crime in general and is rarely problematized in public discourse. The period we live in can be characterized as a period in which a "new punitiveness" has emerged, reflected in "values, strategies, language and practices" (Pratt et al., 2005, p. xi). Behind the new punitiveness are strategies that emphasize incapacitation (over rehabilitation) as the main motivation for prison sentences (Moore & Hannah-Moffat, 2005, p. 97). The simple theory behind the idea that "prison works"—as the slogan by the British Conservative Party goes—is that it incapacitates "criminals" and hence prevents them from committing further crimes. The era of mass incarceration and the new punitive turn hence rests on the control over movement in space, on *immobilization*. To be incarcerated is to be stripped off any control or autonomy not only over how one is doing time but also how one is "doing space." This is a general and very basic idea that has been underpinning penal regimes in the Western world at least since the inception of the modern prison, as we discuss in this book.

Prison media, however, introduces an ambiguity to the idea of immobilization. For one thing, the introduction of new media technologies opened the way for greater mobility within the prison. CCTV and culvert systems allowed incarcerated individuals to move around within the prison parameters more freely than before. Their movements could be tracked, monitored, and steered through automated or semiautomated systems for surveillance. This also followed a more general paradigm shift in how communication was understood in the prison context. In the early modern cell prisons, communication between incarcerated individuals and the staff was strictly prohibited. Silence and isolation were so important that they were reinforced in the first paragraph of the handbook for prison wardens published in the mid-nineteenth century in Sweden. The architecture of the cell prison as well as the interior design and various props and technologies used within the facilities (prison masks, cell cabinets, etc.) all enhanced and underlined this mandatory noncommunication. The communicative regime of the cell prison was one of centrally controlled communication (sermons, teaching) and hence mirrors ideas of mass communication and broadcasting emerging in the shift of the nineteenth and twentieth centuries. The modern cell prison is a media technology for imparting messages to receivers in a clearly defined space. The bunker prisons of the 1950s and 1960s in this respect represent a different communicative ideal: communication between incarcerated individuals and with the staff is now the ideal. Communication becomes increasingly mandatory as part of the rehabilitation efforts. The well-behaved incarcerated individual is one who communicates, and the architecture and interior design is staged to facilitate and enhance such communication: communal areas, spaces for movement and transport, and different media technologies are introduced within the prison, and the incarcerated individuals are encouraged to participate in art and media projects. To engage in communication is to be on the road to reform. Space was changed to become a vessel for movement and communication, but the psychical space of the prisons themselves was also transformed through these communication and art projects, such as the mural paintings done by incarcerated individuals.

But what about our present moment and of the future? Zygmunt Bauman (1998) has stated that immobilization is the logically perfect punishment of our time, a time in which an individual's sense of being free is intertwined with a right to personal mobility. In that sense the punitive turn in contemporary society and that of prisons as a "logical" or "natural" form of punishment should come as no surprise. To be prohibited from moving is a far more potent punishment than what it might have been in the nineteenth-century context, in which the individuals' movement was more restricted also outside of the prison context.

Paradoxically, however, the increased mobility of incarcerated individuals, instigated in the prisons of the 1950s and 1960s, has intensified in new ways through new forms of prison media in contemporary society. From the 1990s onward, as we discussed in chapter 5, ankle monitors have been introduced as a way of dispersing incarceration beyond the physical confinement of the prison itself. This technology has extended the logics of the prison to the world outside the walls in what can be called "mobile incarceration."

Electronic monitors, especially when combined with GPS and other forms of tracking, is a way of extending the punitive logics outside the prison. This represents control and surveillance of movement while providing a certain degree of freedom and autonomy of movement to the incarcerated. This "freedom" is in many ways illusionary, and the range and intensity of control that these technologies represent might well exceed previous forms of control and surveillance of incarcerated individuals. Electronic monitoring has nevertheless in many jurisdictions been seen and discursively framed as "the future" of incarceration. Often it has been introduced without much public debate. The fact that the authorities see this form of punishment as a promise for the future is in part due to the fact that it is far less costly than imprisonment, but it also is connected to rehabilitative potentials. Electronic monitoring devices for tracking offenders have furthermore been remediated and popularized not only in the form of surveillance devices (such as for children) but also activity tracking devices for self-monitoring (such as ankle-worn trackers and smartwatches). These technologies then demonstrate the crucial and direct relationship between prison technologies and technologies developed for communication and control.

Mobile incarceration is not restricted to the prison and to penal contexts as such. Rather, it can be seen as an emerging cultural and social logic of contemporary societies. The discourses of electronic monitoring are, as has been argued by Nellis (2005), "primarily grounded in, and expressive of, the managerial-surveillant enterprise—it facilitates the efficient, fine-grained regulation of human action in time and space" (p. 181). And this enterprise stretches far beyond the prison walls and the context of punishment. On the contrary, mobile incarceration is something that permeates society more generally, as we are all immersed in networks, systems, and technologies that render us locatable and traceable. Nellis (2005) argues that "our willingness to be watched by ubiquitous CCTV systems, our collective immersion in telecommunication and financial networks, our participation in managerialized workplaces mean that we are all somewhat locatable and traceable, and by and large we are seduced by the narratives that present this as convenience, reassurance, and pleasure" (p. 181).

The penal logic is dispersed into other institutions and social spheres through the export of technologies that have been developed for surveillance and punishment. Ankle monitoring as well as other forms of electronic monitoring are also an example of how ideas, cultural logics, and technologies connecting the outside world and the prison are emerging. They make visible the underlying dialectical relation between the carceral condition and society. Through discourses of efficiency and convenience, digital technologies for monitoring and control (mobile phones, social media, etc.) are being used on a large scale by the public. This general social development reinforces the discourse of electronic monitoring as a "solution" to challenges of the corrections system. Contemporary digital culture that is largely based on digital media for monitoring and surveillance and the increased use of smart surveillance technology in corrections together constitute the contemporary moment of mobile incarceration.

Mobile incarceration is, then, a web of discourses in which surveillance, control, and loss of spatial autonomy are imagined and described as gaining possibilities for freedom and mobility. Raymond Williams (1974/1990) proposed the term "mobile privatization" to capture the role of electronic mass media in the twentieth century. For him, television was a mediator between contradictory dominant modern ideals of mobility on the one hand and home-centered living on the other. Mobile incarceration, correspondingly, suggests a mediation between the paradoxes of our contemporary moment, between the praised right to private mobility and our willingness to be watched by and immersed in networks of control and surveillance, a willingness that is accomplished through the seduction "by narratives that present this as convenience, reassurance, and pleasure" (Nellis, 2005, p. 181).

The future of prison media in this sense might take place largely outside the prison itself, extending the punitive principles of control and surveillance to society at large. And in that sense, prison media is a magnifying glass that allows us to see and scrutinize imaginaries and ideas of the future of our societies at large.

PRISON MEDIA AS THE FOUNDATION BOLT

The media work conducted in prisons, the mediating role of prison architecture, and the media technologies designed and developed in the prison bring to the fore certain tensions and paradoxes of media more generally. Media carry the idea of technologies of freedom (Pool, 1983). They carry the potential of and contribute to the possibility of self-expression and creativity. However, media are always also produced in the context of deprivation of freedom. Prison media manifest this ambiguous position of media more generally. Prison media illustrate how media extend the human capacities for communication and at the same time maintain and control them. Most importantly, prison media underpin important dimensions of everyday life beyond the prison walls, even as their emergence, the work that is producing them, remains largely hidden from public attention. Incarcerated individuals have in important ways contributed to our media infrastructures and continue to do so, be it through digital piecework such as scanning newspapers and public records or producing parts of the media hardware.

Prison media form in many ways the foundation bolt of prisons, and if you "find the foundation bolt and remove it, the prison collapses" (Ahl, 1974, p. 9). Prison media hold the foundation of places of incarceration. Without them the penal regime collapses in its current form. The novel *Grundbulten* (The foundation bolt) ends with a reflection on the boundlessness of incarceration: "We know something now that we did not know before. Deprived of the big picture view, we imagined that there is a boundary, a real and concrete boundary of incarceration, or oppression. But now we have understood that there aren't such boundaries and that they will not emerge soon. So next time, we know what we are not going to do. And maybe we kind of feel what we should be doing. Next time. And next time" (Ahl, 1974, p. 251).

What are the boundaries of prison media? Understood as technologies, prison media have been imagined as ways to organize penal institutions more efficiently. Not only with the contemporary smart prison and digital technology but already in the 1950s with CCTV technology, prison and probation services saw a potential in technology to reduce costs through the automation of surveillance. Modern technology maintains ideas of rationalization and efficiency, and we ought to ask what the boundaries of rationalization and efficiency are with the help of technology in the prison context and beyond.

As Jack Qiu (2016) argues, our media technologies have been developed based on an exploitative structure that preserves fundamental inequalities in society and on a global scale. Even if Swedish and other jurisdictions have seen that mobilizations for incarcerated people's rights and their everyday lives have been improved over time, this group has remained in its subordinate position, being defined as deviant from the norm and legality. We, however, argue that these technologies are essential to our media infrastructures in different ways. This perspective allows us to account for the importance of subaltern groups for media development more generally while they have only limited agency in that process. If our media systems have been produced, maintained, and developed in heterotopian contexts by subaltern groups who are fundamentally deprived of freedom and certain civic rights, we are called to rethink the role and aura that media carry in our societies, an aura of freedom and possibility, of creativity and inspiration, that is increasingly situated at the heart of social power (Couldry, 2000). To fully understand the social role and power of media, we need to consider their debt to the prison. In turn, prisons as societal institutions emerge in the context of specific media cultures and are hence also a product of the available media technologies as well as discourses. While prisons play an important role for our media infrastructures, media infrastructures play an essential role for how prisons take shape.

NOTES

PREFACE

1. BBC News (2019, August 14), *ASAP Rocky: A complete timeline of rapper's assault case*, https://www.bbc.com/news/newsbeat-49333962

CHAPTER 1

1. Kristin Houser (2019, February 14), *Hong Kong has a plan to make all of its prisons "smart,"* Futurism, https://futurism.com/smart-prisons-hong-kong

2. Singapore Ministry of Home Affairs (n.d.), *A prison without guards: Where technology enhances operational effectiveness*, https://www.mha.gov.sg/home-team-news/story/detail/a-prison-without-guards-where-technology-enhances-operational-effectiveness/

3. Criminal Sanctions Agency (n.d.), *German Gerdes selected as the smart prison system supplier*, https://www.rikosseuraamus.fi/en/index/topical/pressreleasesandnews/Pressreleasesandnews2020/germangerdesselectedassmartprisonsystemsupplier.html

4. Linda Snecker (2014), FN-kritik mot långa häktningstider och omfattande restriktioner [UN criticism against long detention times and extensive restrictions], Sveriges Riksdag, https://www.riksdagen.se/sv/webb-tv/video/interpellationsdebatt/fn-kritik-mot-langa-haktningstider-och-omfattande-restriktioner_H210172; SVT Nyheter (2019, July 24), *Ingen förändring trots svidande häkteskritik* [No change despite stinging prison criticism], https://www.svt.se/nyheter/inrikes/ingen-forandring-trots-svidande-hakteskritik

CHAPTER 3

1. *Hallbladet*: 3 (1963, pp. 6–7).

2. Kungl. Maj.ts proposition nr 166 år 1957, n.p.

3. Kungl. Maj.ts proposition nr 166 år 1957, n.p.

4. Kommittén för anstaltsbehandling inom Kriminalvården, PM nr 15 30.8.68.

5. Svenska transportarbetareförbudet 1954–1970. Arbetarrörelsens Arkiv och Bibliotek, volym 1823/F/1/108/2.

6. Kriminalvården, Press release, May 3, 2018, https://www.kriminalvarden.se/globalassets/kontakt_press/pressmeddelanden/krimtech-klar.pdf

7. Guardian RFID (n.d.), *Overview*, https://guardianrfid.com/about

8. Camille Knighton (2018, July 19), *Why jails should switch to mobile inmate tracking*, Guardian RFID, https://guardianrfid.com/blog/why-jails-should-switch-to-mobile-inmate-tracking

9. From promotional material by GTL.

CHAPTER 4

1. The prisons opened in Sweden in this period were: Hinseberg (1956), Skogome (1957), Hällby (1958), Mariefred (1958), Norrtälje (1959), Tidaholm (1959), Tillberga (1963), Svartsjö (1965), Asptuna (1965), Kumla (1965), Skänninge (1966), and Österåker (1969).

2. When the City Prison in Stockholm opened in 1852, for example, each cell had a water closet. It would take 17 more years before a water closet was installed in a private home in Sweden (Åman, 1976, p. 110).

3. *Hallbladet*: 1–2 (1977, p. 15).

CHAPTER 5

1. Telio brochure distributed at the expo in Lisbon.

2. *Göteborgsposten*, September 5, 1993.

3. *Dagens Nyheter*, January 16, 1992.

4. iDag, January 31, 1995.

5. *Göteborgsposten*, May 9, 1993.

6. *Dagens Nyheter* October 2, 1995; *Dagens Nyheter*, January 21, 1993.

7. Sveriges Riksdag (n.d.), *Lag (1994:451) om intensivövervakning med elektronisk kontroll*, https://www.riksdagen.se/sv/dokument-lagar/dokument/svensk-forfattningssamling/lag-1994451-om-intensivovervakning-med_sfs-1994-451

8. Sveriges Radio (2021, April 21), *Fotbojor på flera anstalter har slutat fungera*, https://sverigesradio.se/artikel/fotbojor-pa-flera-anstalter-har-slutat-fungera

9. Sveriges Riksdag (n.d.). *Användning av fotboja i kriminalvården*, https://www.riksdagen.se/sv/dokument-lagar/dokument/motion/anvandning-av-fotboja-i-kriminalvarden_GW02Ju389

10. KVFS 2016:6 - Kriminalvårdens föreskrifter om försöksverksamhet med GPS-övervakning vid permission och särskilda utslussningsåtgärder, https://www.kriminalvarden.se/om-kriminalvarden/publikationer/foreskrifter/kvfs-20166---kriminalvar

dens-foreskrifter-om-forsoksverksamhet-med-gps-overvakning-vid-permission-och-sarskilda-utslussningsatgarder/

11. Regeringskansliet (2020, November 16), *Effektivare förfarande och utökad kontroll vid verkställighet av fängelsestraff med fotboja*, https://www.regeringen.se/rattsliga-dokument/departementsserien-och-promemorior/2020/11/effektivare-forfarande-och-utokad-kontroll-vid-verkstallighet-av-fangelsestraff-med-fotboja/

12. Philips (2020, May 26), *Philips launches next generation wearable biosensor for early patient deterioration detection, including clinical surveillance for COVID-19*, https://www.philips.com/a-w/about/news/archive/standard/news/press/2020/20200526-philips-launches-next-generation-wearable-biosensor-for-early-patient-deterioration-detection-including-clinical-surveillance-for-covid-19.html-

13. Catalog of the 38th edition of the American Jail Association expo, 2019.

14. Presentation by Håkan Klarin.

15. https://www.vinnova.se/en/m/testbed-sweden/testbeds-in-sweden/

16. https://www.media.mit.edu/publications/a-living-laboratory-for-the-design-and-evaluation-of-ubiquitous-computing-technologies-2/

17. Skywalk Labs (n.d.), *We build products to radically improve quality of life in cities for all*, https://www.sidewalklabs.com/

18. Testbädd Sverige (n.d.), *Vinnova*, https://www.vinnova.se/m/testbadd-sverige/

REFERENCES

Adamson, M. (2004). *När botten stack upp: Om de utslagnas kamp för frihet ochmänniskovärde* [When the bottom of the stack fought back: The struggle of the excluded for freedom and human dignity]. Hedemora: Gidlund.

Ahl, K. (1974). *Grundbulten* [The foundation bolt]. Stockholm: Prisma.

Allard, T., Wortley, R., & Stewart, A. (2006). The purposes of CCTV in prison. *Security Journal, 19*, 58–70.

Andersen, H. C. (1964). *Resa i Sverige* [Travels in Sweden]. Stockholm: Rabén & Sjögren.

Anderson, B. (2010). Preemption, precaution, preparedness: Anticipatory action and future geographies. *Progress in Human Geography, 34*(6), 777–798. doi:10.1177/0309132510362600

Andrejevic, M. (2002). The work of being watched: Interactive media and the exploitation of self-disclosure. *Critical Studies in Media Communication, 19*(2), 230–248.

Andrejevic, M. (2007). *iSpy: Surveillance and power in the interactive era*. Lawrence: University Press of Kansas.

Appadurai, A., & Alexander, N. (2020). *Failure*. Cambridge, UK: Polity.

Appel, H., Anand, N., & Gupta, A. (2018). Introduction: Temporality, politics, and the promise of infrastructure. In N. Anand, A. Gupta, & H. Appel (Eds.), *The promise of infrastructure* (pp. 1–38). Durham, NC: Duke University Press.

Appelgren, B. (1967). On the inside you lose all human value [Här inne mister man allt mänskligt värde]. *Aftonbladet*.

Arbetartidningen. (1959). *16 kameror följer fången hela tiden* [16 cameras are following incarcerated constantly]. November 30.

Armstrong, S. (2018). The cell and the corridor: Imprisonment as waiting, and waiting as mobile. *Time & Society, 27*(2), 133–154. doi:10.1177/0961463x15587835

Åman, A. (1976). *Om den offentliga vården: Byggnader och verksamheter vid svenska vårdinstitutioner under 1800- och 1900-talen; En arkitekturhistorisk undersökning* [*On public care: Buildings and activities at Swedish care institutions between 1800 and 1900; An architecture historical study*]. Stockholm: Liber.

Bagaric, M., Hunter, D., & Wolf, G. (2018). Technological incarceration and the end of the prison crisis. *Journal of Criminal Law and Criminology, 108*(1), 73–135.

Baran, P. A., & Sweezy, P. M. (1966). *Monopoly capital: An essay on the American economic and social order*. New York: Monthly Review Press.

Bauman, Z. (1998). *Globalization: The human consequences*. London: Polity.

BBC News. (2019, August 14). *ASAP Rocky: A complete timeline of rapper's assault case*, https://www.bbc.com/news/newsbeat-49333962

Benjamin, R. (2019a). *Race after technology*. Cambridge, UK: Polity.

Benjamin, R. (2019b). Introduction: Discriminatory design, liberating imagination. In R. Benjamin (Ed.), *Captivating technology: Race, carceral technoscience, and liberatory imagination in everyday life* (pp. 1–22). Durham, NC: Duke University Press.

Bentham, J. (1791/1995). *The panopticon writings*. New York: Verso.

Bergsten, U. (1983). Blixten slår ned på Kumla—då är det dags för televerkstan att rycka ut. [Lightning strikes Kumla—then it's time for the teleworkshop to get active]. Kriminalvårdens personaltidning.

Bernard, A. (2019). *The triumph of profiling: The self in digital culture*. Cambridge, UK: Polity.

Blomqvist, H., & Waldetoft, D. (1997). *Zimmermans hotell: Linköpings länscellfängelse, 1846–1946* [Zimmeran's hotel: The cell prison in Linköping 1846–1946]. Stockholm: Nordiska museet.

Blumstein, A. (2011). Bringing down the US prison population. *Prison Journal, 91*, 12S–26S.

Bowden, T. S. (2002). A snapshot of state prison libraries with a focus on technology. *Behavioral & Social Sciences Librarian, 21*(2), 1–12.

Bowker, G., & Star, S. L. (1999). *Sorting things out: Classification and its consequences*. Cambridge, MA: MIT Press.

Brodie, P., & Velkova, J. (2021). Cloud ruins: Ericsson's Vaudreuil-Dorion data centre and infrastructural abandonment. *Information, Communication & Society, 24*(6), 869–885.

Brottsförebyggande rådet. Arbetsgruppen rörande kriminalpolitik [National Swedish Council for Crime Prevention. Working group for criminal policy]. (1977). *Nytt straffsystem: Idéer och förslag* [*A new penal system: Ideas and proposals*]. Stockholm: LiberFörlag.

Brown, M. (2009). *The culture of punishment: Prison, society, and spectacle*. New York: New York University Press.

Browne, S. (2015). *Dark matters: On the surveillance of blackness*. Durham, NC: Duke University Press.

Brunton, F. (2013). *Spam: A shadow history of the internet*. Cambridge, MA: MIT Press.

Bucher, T. (2016). Neither black nor box: Ways of knowing algorithms. In S. Kubitschko & A. Kaun (Eds.), *Innovative methods in media and communication research* (pp. 81–98). Basingstoke: Palgrave Macmillan.

Bunner, T. (1967). Svensk kriminalvård av idag [Swedish corrections today]. *Arkitektur, 67*(11), 640–649.

Carrabine, E. (2008). *Crime, culture and the media*. London: Polity.

Cavadino, M., & Dignan, J. (2006). *Penal systems: A comparative approach*. London: Sage.

Cheney-Lippold, J. (2017). *We are data: Algorithms and the making of our digital selves*. New York: New York University Press.

Clemmer, D. (1940). *The prison community*. Boston: Christopher Publishing House.

Cohen, S., & Taylor, L. (1972). *Psychological survival: The experience of long-term imprisonment*. Harmondsworth, UK: Penguin Books.

Cohen, S., & Taylor, L. (1976). *Escape attempts: The theory and practice of resistance to everyday life*. London: Routledge.

Coleman, R. (2018). Theorizing the present: Digital media, pre-emergence and infra-structures of feeling. *Cultural Studies, 32*(4), 600–622. doi:10.1080/09502386.2017.1413121

Colomina, B. (1994). *Privacy and publicity: Modern architecture as mass media*. Cambridge, MA: MIT Press.

Conrad, S. (2012). Collection development and circulation policies in prison libraries: An exploratory survey of librarians in US correctional institutions. *Library Quarterly, 82*(4), 407–427.

Cornfeld, L. (2020). Videotape and vibrators: An industry history of technosexuality. *Feminist Media Histories, 6*(4), 94–120.

Couldry, N. (2000). *The place of media power: Pilgrims and witnesses of the media age*. London: Routledge.

Couldry, N. (2012). *Media, society, world: Social theory and digital media practice*. Cambridge, MA: Polity.

Council of Europe. (2021). Report to the Swedish Government on the visit to Sweden carried out by the European Committee for the Prevention of Torture and Inhuman or Degrading Treatment or Punishment (CPT), September 9. https://rm.coe.int/1680a3c256

Crawford, K. (2021). *Atlas of AI: Power, politics, and the planetary costs of artificial intelligence*. New Haven, CT: Yale University Press.

Criminal Sanctions Agency. (n.d.). *German Gerdes selected as the smart prison system supplier.* https://www.rikosseuraamus.fi/en/index/topical/pressreleasesandnews/Pressreleasesandnews2020/germangerdesselectedassmartprisonsystemsupplier.html

De Giorgi, A. (2006). *Re-thinking the political economy of punishment: Perspectives on post-Fordism and penal politics.* Aldershot, UK: Ashgate.

Delfanti, A. (2019). Machinic dispossession and augmented despotism: Digital work in an Amazon warehouse. *New Media & Society, 23*(1), 39–55. doi:https://doi.org/10.1177/1461444819891613

Delgoda, J. P. (1980). Prison without guards. *International Journal of Offender Therapy and Comparative Criminology, 24*(3), 244–248.

Denis, J., Mongili, A., & Pontille, D. (2015). Maintenance & repair in science and technology studies. *Tecnoscienza: Italian Journal of Science & Technology Studies, 6*(2), 5–15.

Deuze, M. (2007). *Media work.* Cambridge, MA: Polity.

Deuze, M. (2012). *Media life.* Cambridge, MA: Polity.

Doan, C. (2015). *The "subject-effects" of gyms: Studying the interactional, sociospatial and performative order of the fitness site* (PhD thesis). Birkbeck, University of London.

Downey, G. J. (2002). *Telegraph messenger boys: Labor, technology, and geography, 1850–1950.* New York: Routledge.

Dunbar-Hester, C. (2014). *Low power to the people: Pirates, protest, and politics of FM radio activism.* Cambridge, MA: MIT Press.

Edwards, P., Bowker, G., Jackson, S., & Williams, R. (2009). Introduction: An agenda for infrastructure studies. *Journal of the Association for Information Systems, 10*(5), 365–374.

Ekblad, G. (1967). *Centralanstalten i Kumla: Enkät till intagna och personal* [Central prison in Kumla: Survey of incarcerated and staff]. Stockholm: Arbetshandling från byggforskningen.

Ellcessor, E. (2021). The care and feeding of 9-1-1 infrastructure: Dispatcher culture as media work and infrastructural transformation. *Cultural Studies, 35*(4–5), 792–813. DOI: 10.1080/09502386.2021.1895249

Engblom, L.Å. (2020). När tidningsbuden var ett konkurrensmedel [When the newspaper carriers were important for competition between newspapers]. *Mellan det lokala och hyperglobala: Journalistikens förändringar och utmaningar i en digital tid* [*Between the local and hyper-global: Changes in and challenges for journalism in a digital era*], 123–131. Södertörn University: Huddinge.

Engels, F., & Rogge, J.-C. (2018). Tensions and trade-offs in real-world laboratories—the participants' perspective. *GAIA—Ecological Perspectives for Science and Society, 27*(1), 28–31. doi:http://dx.doi.org/10.14512/gaia.27.S1.8

Ericson, S., & Riegert, K. (Eds.). (2010). *Media houses: Architecture, media and the production of centrality.* New York: Peter Lang.

Ericsson. (1976). Verksamhetsberättelse [Annual report]. Stockholm. https://www.ericsson.com/48f7ac/assets/local/investors/documents/financial-reports-and-filings/annual-reports/annual-report_1976_se.pdf

Eriksson, J. (1966). Program för reformer inom det nuvarande anstaltssystemet: Anförande vid Tjyvriksdagen i Strömsund 1966 [Programme for reforms within the current prison system: Speech at the Thief Parliament i Strömsund 1966]. *Liberal Debatt, 19*(6), 6–11.

Eriksson, T. (1939). *Brott och samhälle* [Crime and society]. Stockholm: Bonnier.

Eriksson, T. (1966). Kriminalvård på anstalt [Corrections in the institution]. In J. Freese (Ed.), *Kriminaliteten och samhället* [*Criminality and society*]. Stockholm: Bokförlaget Aldus/Bonniers.

Eriksson, T. (1967). *Kriminalvård: Idéer och experiment*. Stockholm: Norstedt.

Eubanks, V. (2017). *Automating inequality: How high-tech tools profile, police, and punish the poor*. New York: St. Martin's.

Fångvårdens byggnadskommitté. (1956). *Anstalter i Tidaholm, Norrtälje, Säve, Hälsingborg* [Facilities i Tidaholm, Norrtälje, Säve, Hälsingborg]. Stockholm: Justitiedepartementet.

Fångvårdens byggnadskommitté. (1965). *Förslag till utvecklingsplan för kriminalvårdens anstaltssystem 1966–1975* [Proposal for a development plan for corrections facility system 1966–1975]. Stockholm: Justitiedepartementet.

Fångvårdens byggnadskommitté. (1972). *Fångvårdens byggnadskommitté 1956–1972: Slutrapport* [Corrections building committee 1956–1972: Final report]. Stockholm: Justitiedepartementet.

Fångvårdens byggnadskommittés arkiv. (1964). *Relationsritningar till slutna centralanstalten i Kumla* [Blueprints for high security prison in Kumla]. *Riksarkivet*, 20:1755/J. Riksarkivet.

Fassin, D. (2017). *Prison worlds: An ethnography of the carceral condition*. Malden, MA: Polity.

Forsler, I. (2020). *Enabling media: Infrastructures, imaginaries, and cultural techniques in Swedish and Estonian visual arts education*. Huddinge, Sweden: Södertörns högskola.

Forsythe, W. J. (1987). *The reform of prisoners, 1830–1900*. London: Croom Helm.

Foster, R. (2019). "Doing the wait": An exploration into the waiting experiences of prisoners' families. *Time & Society*, 28(2), 459–477.

Foucault, M. (1975). *Surveiller et punir: Naissance de la prison*. Paris: Gallimard.

Foucault, M. (1979). *Discipline and punish: The birth of the prison*. London: Penguin Books.

Foucault, M. (1984). Of other spaces: Utopias and heterotopias. *diacritics, 16*(1), 22–27.

Fredrikzon, J. (2021). *Kretslopp av data: Miljö, befolkning, förvaltning och den tidiga digitaliseringens kulturtekniker* [*Cycles of data: Environment, population, administration and the cultural techniques of early digitalization*]. Lund: Mediehistoriskt Arkiv.

Fuchs, C. (2009). Social networking sites and the surveillance society: A critical case study of the usage of StudiVZ, Facebook, and MySpace by students in Salzburg in the context of electronic surveillance. http://fuchs.uti.at/wp-content/uploads/SNS_Surveillance_Fuchs.pdf

Fuchs, C. (2014). Digital prosumption labour on social media in the context of the capitalist regime of time. *Time & Society, 23*(1), 97–123. doi:10.1177/0961463X13502117

Gable, R. S., & Gable, R. K. (2016). Remaking the electronic tracking of offenders into a "persuasive technology." *Journal of Technology in Human Services 34*(1), 13–31. doi:10.1080/15228835.2016.1138839

Gahrton, P. (1971). Kumla-serien—en studie i organiserat oförnuft [Kumla series: A study of organised stupidity]. *Pockettidningen R, 1*, 78–86.

Garland, D. (2001). *The culture of control: Crime and social order in contemporary society*. Chicago: Chicago University Press.

Gehl, R. (2014). *Reverse engineering social media: Software, culture, and political economy in new media capitalism*. Philadelphia: Temple University Press.

Gill, R. (2002). Cool, creative and egalitarian? Exploring gender in project-based new media work in Euro. *Information, Communication & Society, 5*(1), 70–89.

Gilmore, R. W. (2007). *Golden gulag: Prisons, surplus, crisis and opposition in globalizing California*. Berkeley: University of California Press.

Goffman, E. (1961). *Asylums: Essays on the social situation of mental patients and other inmates*: Anchor Books.

Goldberg, E., & Evans, L. (2009). *The prison industrial complex and the global economy*. Oakland, CA: PM Press.

Gray, M. L., & Suri, S. (2019). *Ghost work: How to stop Silicon Valley from building a new global underclass*. Boston: Houghton Mifflin.

Greenberg, J. (2008). *From Betamax to Blockbuster: Video stores and the invention of movies on video*. Cambridge, MA: MIT Press.

Grip, L., Caviezel, S., & Öjes, E. (2018). *How architecture and design matter for the Swedish Prison and Probation Services: A rapid review of the literature*. Norrköping: The Research and Evaluation Unit, Swedish Prison and Probation Services.

Gustavsson, G. (1989). *Det nya cellfängelset i Falun 1848 och tre av dess första fångar: Bakgrunden* [The new prison in Falun 1848 and three of its first incarcerated: Background]. Falun: Högskolan i Falun/Borlänge.

Hällby Prison. (1988). *Annual Report 1987–1988*.

Hancock, P., & Jewkes, Y. (2011). Architectures of incarceration: The spatial pains of imprisonment. *Punishment & Society, 13*(5), 611–629.

Hemmungs, Wirtén, E. (2019). How patents became documents, or dreaming of technoscientific order, 1895–1937. *Journal of Documentation, 75*(3), 577–592.

Hepp, A. (2016). Pioneer communities: Collective actors in deep mediatisation. *Media, Culture & Society, 38*(6), 918–933.

Herrity, K. (2020). Hearing behind the door: The cell as a portal to prison life. In J. Turner & V. Knight. (Eds.), *The prison cell: Embodied and everyday spaces of incarceration* (pp. 239–260). Cham, Germany: Palgrave Macmillan.

Hesmondhalgh, D., & Baker, S. (2011). *Creative labour: Media work in the three cultural industries*. London: Routledge.

Hibbert, C. (1963). *The roots of evil: A social history of crime and punishment*. London: Weidenfeld & Nicolson.

Hjelm, C. (1848). Hr majoren Hjelms berättelse om fängelsebyggnaderne i England, Frankrike och Tyskland [Major Hjelm's account of the prison buildings in England, France and Germany]. *Aftonbladet (9)*13.

Hobsbawm, E. J. (1977). *The age of revolution: Europe, 1789–1848*. London: Abacus.

Hockenhull, M., & Cohen, M. L. (2021). Hot air and corporate sociotechnical imaginaries: Performing and translating digital futures in the Danish tech scene. *New Media & Society, 23*(2), 302–321. doi:10.1177/1461444820929319

Hogan, M. (2015). Data flows and water woes: The Utah data centre. *Big Data & Society, 2*(2), 1–12. doi:10.1177/2053951715592429

Holt, J., & Vonderau, P. (2015). "Where the internet lives": Data centers as cloud infrastructure. In L. Parks & N. Starosielski (Eds.), *Signal traffic: Critical studies of media infrastructures* (pp. 71–94). Urbana: University of Illinois Press.

Houser, K. (2019). Hong Kong has a plan to make all of its prisons "smart." *Futurism*, February 15. https://futurism.com/smart-prisons-hong-kong

Howard, J. (1777). *The state of the prisons in England and Wales, with preliminary observations, and an account of some foreign prisons*. Warrington, UK: William Eyres.

Ihsgren, I. (1974). *Sociopaten*. Stockholm: Prisma.

Irani, L. (2015). The cultural work of microwork. *New Media & Society, 17*(5), 720–739.

Jackson, S. (2014). Rethinking repair. In T. Gillespie, P. Boczkowski, & K. Foot (Eds.), *Media technologies: Essays on communication, materiality and society*. Cambridge, MA: MIT Press.

Jaffe, A. B., & de Rassenfosse, G. (2019). Patent citation data in social science research: Patent citation data in social science research; Overview and best practices. *Journal of the Association for Information Science and Technology, 68*(6), 1360–1374.

Jakobsson, P., & Stiernstedt, F. (2010). Pirates of Silicon Valley: State of exception and dispossession in Web 2.0. *First Monday, 15*(7). doi:https://doi.org/10.5210/fm.v15i7.2799

Jakobsson, P., & Stiernstedt, F. (2012). Time, space and clouds of information: Data centre discourse and the meaning of durability. In G. Bolin (Ed.), *Cultural technologies* (pp. 115–129). London: Routledge.

Jareborg, N. (1995). Vilken sorts straffrätt vill vi ha? Om defensiv och offensiv straffrättspolitik [What kind of criminal law do we want? On defensive and offensive criminal law policy]. In D. Victor & A. Ashworth (Eds.), *Varning för straff: Om vådan av den nyttiga straffrätten*. Stockholm: Fritze.

Järveläinen, E., & Rantanen, T. (2020). Incarcerated people's challenges for digital inclusion in Finnish prisons. *Nordic Journal of Criminology*, 1–21. doi:10.1080/2578983X.2020.1819092

Jarzombek, M. (2010). Corridor space. *Critical Inquiry, 36*(4), 728–770.

Jasanoff, S. (2005). *Designs on nature: Science and democracy in Europe and the United States*. Princeton, NJ: Princeton University Press.

Jasanoff, S. (2015). Future imperfect: Science, technology, and the imaginations of modernity. In S. Jasanoff & S.-H. Kim (Eds.), *Dreamscapes of modernity: Sociotechnical imaginaries and the fabrication of power* (pp. 1–33). Chicago: University of Chicago Press.

Jasanoff, S., & Kim, S.-H. (2009). Containing the atom: Sociotechnical imaginaries and nuclear regulation in the U.S. and South Korea. *Minerva, 47*(2), 119–146.

Jefferson, B. (2020). *Digitize and punish: Racial criminalization in the digital age*. Minneapolis: University of Minnesota Press.

Jewkes, Y. (2002). The use of media in constructing identities in the masculine environment of men's prisons. *European Journal of Communication, 17*(2), 205–225.

Jewkes, Y., & Johnston, H. (2009). Cavemen in an era of speed-of-light technology: Historical and contemporary perspectives on communication within prisons. *Howard Journal, 48*(2), 132–143.

Jewkes, Y., & Reisdorf, B. (2016). A brave new world: The problems and opportunities presented by new media technologies in prisons. *Criminology & Criminal Justice, 16*(5), 534–551.

Jhally, S., & Livant, B. (1986). Watching as work: The valorisation of audience consciousness. *Journal of Communication, 36*(3), 124–143.

Johnston, H. (2020). "The solitude of the cell": Cellular confinement in the emergence of the modern prison, 1850–1930. In J. Turner & V. Knight (Eds.), *The prison cell: Embodied and everyday spaces of incarceration* (pp. 23–44). Cham: Palgrave Macmillan.

Johnston, N. B., Finkel, K., & Cohen, J. (1994). *Eastern state penitentiary: Crucible of good intentions*. Philadelphia: Philadelphia Museum of Art.

Kaun, A., Logsdon, A., Seuferling, P., & Stiernstedt, F. (forthcoming). Serving machines: Technology development in heterotopian media backends. In L. Parks, J. Velkova, & S. de Ridder (Eds.), *Media backends: The politics of infrastructure, clouds, and artificial intelligence*. Champaign: University of Illinois Press.

Kaun, A., & Stiernstedt, F. (2020a). Doing time, the smart way? Temporalities of the smart prison. *New Media & Society, 22*(9), 1580–1599.

Kaun, A., & Stiernstedt, F. (2020b). Prison media work: From manual labor to the work of being tracked. *Media, Culture & Society, 42*(7–8), 1277–1292.

Kaun, A., & Stiernstedt, F. (2022). Prison tech: Imagining the prison as lagging behind *and* as a test bed for technology advancement. *Communication, Culture and Critique, 15*(1), 69–83, *OnlineFirst*, 1–15.

KBS Rapport. (1969). Slutna fångvårdsanstalter: Inventeringar avseende mått, kvalitéer och tekniska lösningar [Closed correctional facilities: Inventories of dimensions, qualities and technical solutions], *14*(September).

Kellokumpu, R. (1997). Tidningsbud [News carriers]. In *Alla dagar alla nätter: Fyra transporthistoriska essäer* [All days all nights: four transport history essays] (pp. 145–191). Stockholm: Svenska transportarbetareförb.

Khair, M. (2018). A prison without guards: Where technology enhances operational effectiveness. *Singapore Prison Services*, July 6. https://www.mha.gov.sg/home-team-news/story/detail/a-prison-without-guards-where-technology-enhances-operational-effectiveness/

Kindgren, J., & Littman, L. (2015). *Arbete, utbildning och behandling i svenska anstalter: En studie om intagnas sysselsättning* [Work, education and treatment in Swedish prisons: A study of inmates' employment]. https://bra.se/publikationer/arkiv/publikationer/2015-11-16-arbete-utbildning-och--behandling-i-svenska-anstalter.html

Kitchin, R. (2014). The real-time city? Big data and smart urbanism. *GeoJournal, 79*(1), 1–14. doi:10.1007/s10708-013-9516-8

Knauff, K. (2012). *Den klassicistiska vändningen i det tidiga 1900-talets svenska arkitektur: En studie av Liljevalchs konsthall, Kungstornen och Kanslihuset i Stockholm* [The classical turn in the early 20th-century Swedish architecture: A study of Liljevalch's Kunsthall, Kungstornen and Kanslihuset in Stockholm]. Stockholm: KTH Royal Institute of Technology.

Knight, V. (2016). *Remote control: Television in prison*. Basingstoke, UK: Palgrave Macmillan.

Knight, V., & Bennett, J. (2020). Reading Bronson from deep on the inside: An exploration of prisoners watching prison films. In *The Palgrave handbook of incarceration in popular culture* (pp. 33–49). Cham, Switzerland: Palgrave Macmillan.

Kotova, A. (2019). "Time . . . lost time": Exploring how partners of long-term prisoners experience the temporal pains of imprisonment. *Time & Society, 28*(2), 478–498.

Kriminalvården. (1983). *Tidning för Kriminalvårdsverkets personal*, March.

Kriminalvårdsstyrelsen. (1995). *Arbetsdriftens vägval: Rapport* [Work operations' future: Report]. Norrköping: Kriminalvårdsstyrelsen.

Kubitschko, S., & Schütz, T. (2016). Humanitarian media intervention: Infrastructuring in times of forced migration. *spheres: Journal for Cultural Design* (3), 1–14.

Lamroth, A. (1990). *Kåkfångare: Fängelselivet på Långholmen, 1946–1975* [Prisoners: Prison life at Långholm, 1946–1975]. Stockholm: Stockholms stad.

Lappi-Seppälä, T. (2007). Penal policy in Scandinavia. *Crime and Justice, 36*(1), 217–295.

Larkin, B. (2008). *Signal and noise: Media, infrastructure, and urban culture in Nigeria.* Durham, NC: Duke University Press.

Larkin, B. (2018). Promising forms: The political aesthetics of infrastructure. In N. Anand, A. Gupta, & H. Appel (Eds.), *The promise of infrastructure* (pp. 175–202). Durham NC: Duke University Press.

Lauritsen, P., & Feuerbach, A. (2015). CCTV in Denmark 1954–1982. *Surveillance & Society*, 13(3/4): 528–538.

Law, J. (2013). Collateral realities. In P. Baert and F. Domínguez Rubio (Eds.), *The politics of knowledge* (pp. 156–178). London: Routledge.

Law, J., & Ruppert, E. (2013). The social life of methods: Devices. *Journal of Cultural Economy, 6*(3), 229–240. doi:10.1080/17530350.2013.812042

Lehtiniemi, T., & Ruckenstein, M. (2022). Prisoners training AI. In M. Berg, D. Lupton, S. Pink, & M. Ruckenstein (Eds.), *Everyday automation*. London: Routledge.

Lindhagen, C. (1929). Motioner i Första kammaren [Motions in the first chamber], Nr 275, 1929, s. 1.

Lobe, A. (2019). *Speichern und strafen: Die gesellschaft im datengefängnis* [*Safe and punish: The society in the data prison*]. Muenchen, Germany: C. H. Beck.

Luckhurst, R. (2019). *Corridors: Passages of modernity*. London: Reaktion Books.

Lundberg, S. (1997). *Härlanda Fängelse—en tidsspegel* [Härlanda Prison—a time mirror]. Göteborg: Tre Böcker.

Lundgren, F. (2003). *Den isolerade medborgaren: Liberalt styre och uppkomsten av det sociala vid 1800-talets mitt* [The isolated citizen: Liberal government and the rise of the social in the mid-19th century]. Hedemora: Uppsala University.

Lundgren, L., & Evans, C. E. (2017). Producing global media memories: Media events and the power dynamics of transnational television history. *European Journal of Cultural Studies, 20*(3), 252–270.

Mager, A., & Katzenbach, C. (2021). Future imaginaries in the making and governing of digital technology: Multiple, contested, commodified. *New Media & Society, 23*(2), 223–236. doi:10.1177/1461444820929321

Magnet, S. (2011). *When biometrics fail: Gender, race, and the technology of identity*. Durham, NC: Duke University Press.

Markus, T. A. (1993). *Buildings & power: Freedom and control in the origin of modern building types*. London: Routledge.

Martin, R. (2003). *The organizational complex: Architecture, media, and corporate space*. Cambridge, MA: MIT Press.

Martinson, R. (1974). What works?—Questions and answers about prison reform. *Public Interest, 35*, 22–54.

Marvin, C. (1988). *When old technologies were new: Thinking about electric communication in the late nineteenth century*. New York: Oxford University Press.

Mathiesen, T. (1974). *The politics of abolition*. New York: Wiley.

Mathiesen, T. (2016). *The politics of abolition revisited*. London: Routledge.

Mattern, S. (2016). Scaffolding, hard and soft—Infrastructures as critical and generative structures. *spheres: Journal for Digital Cultures* (3). https://spheres-journal.org/wp-content/uploads/spheres-3_Mattern.pdf

Matthews, R. (2009). *Doing time: An introduction to the sociology of imprisonment*. Basingstoke, UK: Palgrave Macmillan.

Mayer, V. (2011). *Below the line: Producers and production studies in the new television economy*. Durham, NC: Duke University Press.

Medlicott, D. (1999). Surviving in the time machine: Suicidal prisoners and the pains of prison time. *Time & Society, 8*(2–3), 211–230.

Meisenhelder, T. (1985). An essay on time and phenomenology of imprisonment. *Deviant Behavior, 6*(1), 39–56.

Melossi, D., & Pavarini, M. (1981/2018). *The prison and the factory: Origins of the penitentiary system*. London: Macmillan.

Mentzer, A. (1878). *Handbok för cellfängelse-föreståndare* [Handbook for cell prison managers]. Karlstad.

Meranze, M. (1996). *Laboratories of virtue: Punishment, revolution, and authority in Philadelphia, 1760–1835*. Chapel Hill: University of North Carolina Press.

Meyrowitz, J. (1986). *No sense of place: The impact of electronic media on social behavior*. Oxford: Oxford University Press.

Mitrea, O. S. (2015). Instances of failures, maintenance, and repair in smart driving. *Tecnoscienza: Italian Journal of Science and Technology Studies, 6*(2), 61–80.

Mol, A. (2002). *The body multiple: Ontology in medical practice*. Durham, NC: Duke University Press.

Moore, D., & Hannah-Moffat, K. (2005). The liberal veil: Revisiting Canadian penality. In J. Pratt, D. Brown, M. Brown, S. Hallsworth, & W. Morrison (Eds.), *The new punitiveness* (pp. 111–126). London: Taylor and Francis.

Morozov, E. (2013). *To save everything, click here: The folly of technological solutionism*. New York: Public Affairs.

Musson, A. E. (1974). Newspaper printing in the Industrial Revolution. *Economic History Review, 10*(3), 411–426.

Nellis, M. (2005). Electronic monitoring, satellite tracking, and the new punitiveness in England and Wales. In J. Pratt, D. Brown, M. Brown, S. Hallsworth, & W. Morrison (Eds.), *The new punitiveness* (pp. 167–185). London: Taylor and Francis.

Nellis, M., & Bungerfeldt, J. (2013). Electronic monitoring and probation in Sweden and England and Wales: Comparative policy developments. *Probation Journal, 60*(3), 278–301.

Niitamo, V., Kulkki, S., Eriksson, M., & Hribernik, K. A. (2006). State-of-the-art and good practice in the field of living labs. Paper presented at the 2006 IEEE International Technology Management Conference (ICE), Milan.

Nilsson, R. (2013). *Från cellfängelse till beteendeterapi: Fängelse, kriminalpolitik och vetande, 1930–1980* [From solitary confinement to behavioral therapy: Prison, penal policy and knowledge, 1930–1980]. Malmö: Égalité.

Nilsson, R. (2017). "First we build the factory, then we add the institution": Prison, work and welfare state in Sweden c. 1930–1970. In P. S. Scharff & T. Ugelvik (Eds.), *Scandinavian penal history, culture and practice: Embraced by the welfare state?* (pp. 35–56). London: Palgrave Macmillan.

Norris, C., & Armstrong, G. (1998). *The maximum surveillance society: The rise of CCTV*. Oxford: Berg.

Norris, C., & McCahill, M. (2006). CCTV: Beyond penal modernism? *British Journal of Criminology, 46*(1), 97–118.

Novek, E. (2009). Mass culture and the American taste for prisons. *Peace Review, 21*(3), 376–384.

Ogonowski, C., Ley, B., Hess, J., Wan, L., & Wulf, V. (2013). Designing for the living room: Long-term user involvement in a living lab. Paper presented at CHI '13: CHI Conference on Human Factors in Computing Systems, Paris.

Olsson, P. O. (1967). De aningslösa och de försvarslösa [The unsuspecting and the defenseless]. *Arkitektur, 67*(11), 608–610.

Parks, L. (2005). Cultures in orbit. In *Cultures in orbit*. Durham: Duke University Press.

Parks, L., & Starosielski, N. (2015). Introduction. In L. Parks & N. Starosielski (Eds.), *Signal traffic: Critical studies of media infrastructures* (pp. 1–27). Urbana: University of Illinois Press.

Parks, L., Velkova, J., & Ridder, S. (forthcoming). *Media backends: Digital infrastructures and sociotechnical relations*. Urbana: University of Illinois Press.

Paulsen, M. (1971). Den kriminalpolitiska debatten under de senaste åren [The criminal policy debate in recent years]. PM 29(26/5): 1970. Riksarkivet. Kommittén för anstaltsbehandling inom Kriminalvården (KAIK) [Committee for institutional treatment within the Prison Service (KAIK)]. Vol. 3: 2481.

Peters, J. D. (1999). *Speaking into the air*. Chicago: University of Chicago Press.

Peters, J. D. (2015). *The marvellous clouds: Towards a philosophy of elemental media*. Chicago: University of Chicago Press.

Pool, I. S. (1983). *Technologies of freedom: On free speech in an electronic age*. Cambridge, MA: Harvard University Press.

Pratt, J., D. Brown, S. Hallsworth, M. Brown & W. Morrison (Eds.). (2005). *The new punitiveness: Trends, theories, perspectives*. Cullompton: Willan.

Pratt, J., & Eriksson, A. (2012). In defence of Scandinavian exceptionalism. In T. Ugelvik & J. Dullum (Eds.), *Penal exceptionalism? Nordic prison policy and practice*. London: Routledge.

Preciado, B. (2013). *Testo junkie: Sex, drugs, and biopolitics in the pharmacopornographic era*. New York: Feminist Press.

Qiu, J. (2016). *Goodbye iSlave: A manifesto for digital abolition*. Urbana: University of Illinois Press.

Rahm, L., & Kaun, A. (2022). Imagining mundane automation: Historical trajectories of meaning making around technological change. In S. Pink, M. Berg, D. Lupton, & M. Ruckenstein, M. (Eds.), *Everyday automation: Experiencing and anticipating emerging technologies* (pp. 23-43). London: Routledge.

Reisdorf, B., & Jewkes, Y. (2016). (B)Locked sites: Cases of internet use in three British prisons. *Information, Communication & Society, 19*(6), 771–786.

Reiter, K., Sexton, L., & Sumner, J. (2018). Theoretical and empirical limits of Scandinavian exceptionalism: Isolation and normalization in Danish prisons. *Punishment & Society, 20*(1), 92–112. doi:10.1177/1462474517737273

Riksrevisionsverket. (1979). *Arbetsdriften vid kriminalvårdens anstalter: En granskning av resultatanalys och samordningsmöjligheter vad gäller det yrkesinriktade arbetet* [Work operations of correctional facilities: A review of performance analysis and coordination opportunities in vocational work]. Stockholm: Riksrevisionsverket.

Ringius, G. (1963). *Handbok i bevakningsteknik* [Handbook on surveillance techniques]. Stockholm: Kungliga Fångvårdsstyrelsen.

Ritchie, D. D. (1854). *The voice of our exiles; or stray leaves from a convict ship*. Edinburgh: John Menzies.

Robberechts, J. & Beyens, K. (2020). PrisonCloud: The beating heart of the digital prison cell. In J. Turner & V. Knight (Eds.), *The prison cell: Embodied and everyday spaces of incarceration* (pp. 283–303). Cham, Switzerland: Palgrave Macmillan.

Roberts, S. T. (2019). *Behind the screen*. New Haven, CT: Yale University Press.

Robertson, C. (2021). *The filing cabinet: A vertical history of information*. Minneapolis: University of Minnesota Press.

Rollo. (1966). 435 intagna i Kumla men inte en läkare [435 incarcerated in Kumla but not one single doctor]. *Göteborgsposten, 2*(2).

Ruppert, E., Law, J., & Savage, M. (2013). Reassembling social science methods: The challenge of digital devices. *Theory, Culture & Society, 30*(4), 22–46. https://doi.org/10.1177/0263276413484941

Rusche, G., & Kirchheimer, O. (1939/2009). *Punishment and social structure*. New Brunswick, NJ: Transaction Publishers.

Rusert, B. (2019). Naturalizing coercion: The Tuskegee experiments and the laboratory life of the plantation. In R. Benjamin (Ed.), *Captivating technology: Race, carceral technoscience, and liberatory imagination in everyday life* (pp. 25–49). Durham, NC: Durham University Press.

Sawhney, H., & Lee, S. (2005). Arenas of innovation: Understanding new configurational potentialities of communication technologies. *Media, Culture & Society, 27*(3), 391–414.

Sawyer, W. (2017). How much do incarcerated people earn in each state? Prison Policy Initiative, April 10. https://www.prisonpolicy.org/blog/2017/04/10/wages/

Sawyer, W., & Wagner, P. (2020). Mass incarceration: The whole pie 2020. Prison Policy Initiative, March 14. https://www.prisonpolicy.org/reports/pie2020.html

Scharff Smith, P. (2012a). A critical look at Scandinavian exceptionalism: Welfare state theories, penal populism, and prison conditions in Denmark and Scandinavia. In T. Ugelvik & J. Dullum (Eds.), *Penal exceptionalism? Nordic prison policy and practice*. London: Routledge.

Scharff Smith, P. (2012b). Imprisonment and internet-access: Human rights, the principle of normalisation and the question of access to digital communications technology. *Nordic Journal of Human Rights, 30*(4), 454–482.

Shapiro, A. (2020). "Embodiments of the invention": Patents and urban diagrammatics in the smart city. *Convergence, 26*(4), 751–774. doi:10.1177/1354856520941801

Siegert, B. (2015). *Cultural techniques: Grids, filters, doors, and other articulations of the real*. New York: Fordham University Press.

Singapore Ministry of Home Affairs. (n.d.). *A prison without guards: Where technology enhances operational effectiveness*. https://www.mha.gov.sg/home-team-news/story/detail/a-prison-without-guards-where-technology-enhances-operational-effectiveness/

Smith, P. S., & Ugelvik, T. (Eds.) (2017). *Scandinavian penal history, culture and prison practice: Embraced by the welfare state?* Basingstoke, UK: Palgrave MacMillan.

Snecker, L. (2014). FN-kritik mot långa häktningstider och omfattande restriktioner [UN criticism against long detention times and extensive restrictions]. Sveriges

Riksdag. https://www.riksdagen.se/sv/webb-tv/video/interpellationsdebatt/fn-kritik-mot-langa-haktningstider-och-omfattande-restriktioner_H210172

SOU (Statens offentliga utredningar). (1959). *Fångvårdsanstalters optimala storlek* [The optimal size of correctional facilities]. Stockholm: Justitiedepartmentet.

SOU (Statens offentliga utredningar). (1971). *Kriminalvård i anstalt* [Correctional services within institutions]. Stockholm: Justitiedepartmentet.

Spillers, H. J. (1987). Mama's baby, papa's maybe: An American grammar book. In *The transgender studies reader remix* (pp. 93–104). New York: Routledge.

Star, S. L. (1999). The ethnography of infrastructure. *American Behavioral Scientist, 43*(3), 377–391.

Star, S. L., & Bowker, G. (2002). How to infrastructure. In L. Lievrouw & S. Livingstone (Eds.), *Handbook of new media: Social shaping and consequences of ICTs* (pp. 151–162). London: Sage.

Star, S. L., & Griesemer, J. R. (1989). Institutional ecology, "translations" and boundary objects: Amateurs and professionals in Berkeley's Museum of Vertebrate Zoology, 1907–19. *Social Studies of Science, 19*(3), 387–420.

Star, S. L., & Ruhleder, K. (1996). Steps towards an ecology of infrastructure: Design and access for large information spaces. *Information Systems Research, 7*(1), 111–134.

Star, S. L., & Strauss, A. (1999). Layers of silence, arenas of voice: The ecology of visible and invisible work. *Computer Supported Cooperative Work (CSCW), 8*(1), 9–30.

Starosielski, N. (2015). *The undersea network.* Durham, NC: Duke University Press.

Strathern, M. (1991). *Partial connections.* Savage, MD: Rowman and Littlefield.

Sudbury, J. (2014). *Global lockdown: Race, gender, and the prison-industrial complex.* London: Routledge.

SVT Nyheter. (2019). Ingen förändring trots svidande häkteskritik [No change despite stinging prison criticism], Sveriges television, July 24. https://www.svt.se/nyheter/inrikes/ingen-forandring-trots-svidande-hakteskritik

Swedish Code of Statutes. (1974: 203). *Om kriminalvård i anstalt* [About correctional services in prisons].

Swedish National Council for Crime Prevention. (1999). Intensivövervakning med elektronisk kontroll: Utvärdering av 1997 och 1998 års riksomfattande försöksverksamhet [Intensive electronic monitoring: Evaluation of the 1997 and 1998 nationwide trials]. BRÅ-rapport 1999:4.

Sykes, G. (1958). *The society of captives: A study of a maximum security prison.* Princeton, NJ: Princeton University Press.

Taylor, C. (2003). *Modern social imaginaries.* Durham, NC: Duke University Press.

Teeters, N. K. (1944). *World penal systems: A survey*. Philadelphia: Pennsylvania Prison Society.

Terranova, T. (2000). Free labor: Producing culture for the digital economy. *Social Text, 18*(2), 33–58.

Tham, H. (2001). Law and order as a leftist project? The case of Sweden. *Punishment & Society, 3*(3), 409–426.

Thompson, H. A. (2012). The prison industrial complex: A growth industry in a shrinking economy. *New Labor Forum, 21*(3), 39–47.

Thompson, J. (1995). *Media and modernity: A social theory of the media*. Cambridge, MA: Polity.

Thompson, K., & Zurn, P. (Eds.). (2021). *Intolerable: Writings from Michel Foucault and the Prisons Information Group (1970–1980)*. Minneapolis: University of Minnesota Press.

Tilley, H. (2011). *Africa as living laboratory: Empire, development and the problem of scientific knowledge, 1870–1950*. Chicago: University of Chicago Press.

Tornklint, J. (1971). *Förtroende för fångar* [*Trust in prisoners*]. Stockholm: Bonniers.

Turner, F. (2006). *From counterculture to cyberculture: Stewart Brand, the Whole Earth Network, and the rise of digital utopianism*. Chicago: University of Chicago Press.

Turner, F. (2016). Prototype. In B. Peters (Ed.), *Digital keywords: A vocabulary of information society and culture*. Princeton, NJ: Princeton University Press.

Turow, J. (2021). *The voice catchers: How marketers listen in to exploit your feelings, your privacy, and your wallet*. New Haven, CT: Yale University Press.

Vandebosch, H. (2001). Media use as an adaptation or coping tool in prison. *Communications, 26*(4), 371–388.

Van De Steene, S., & Knight, V. (2017). Digital transformation for prisons: Developing a needs-based strategy. *Probation Journal, 64*(3), 256–268. doi:10.1177/0264550517723722

Velkova, J. (2016). Data that warms: Waste heat, infrastructural convergence and the computation traffic commodity. *Big Data & Society (3)*, 2. doi:2053951716684144

Velkova, J. (2017). *Media technologies in the making: User-driven software and infrastructures for computer graphics production*. Huddinge, Sweden: Södertörn University Press.

Velkova, J., & Kaun, A. (2021). Algorithmic resistance: Media practices and the politics of repair. *Information, Communication & Society, 24*(4), 523–540.

Von Hofer, H. (2003). Prison populations as political constructs: The case of Finland, Holland and Sweden. *Journal of Scandinavian Studies in Criminology and Crime Prevention, 4*(1), 21–38.

Wacquant, L. J. D. (1999). *Les prisons de la misère* [Prisons of poverty]. Paris: Raisons d'agir.

Wahl-Jorgensen, K., & Hanitzsch, T. (2009). *The handbook of journalism studies*. New York: Routledge.

Waldetoft, D. (2005). Fångvården i Nordiska museet [Criminal justice in the Nordic Museum]. In L-E Jönsson & B. Svensson, (Eds.), *I industrisamhällets slagskugga: Om problematiska kulturarv [In the shadow of the industrial society: On contested cultural memory]* (pp. 66–84). Stockholm: Carlssons.

Wang, J. (2018). *Carceral capitalism*. Cambridge, MA: MIT Press.

Weltevrede, E., Helmond, A., & Gerlitz, C. (2014). The politics of real-time: A device perspective on social media platforms and search engines. *Theory, Culture & Society, 31*(6), 125–150. doi:10.1177/0263276414537318

Williams, C. A. (2003). Police surveillance and the emergence of CCTV in the 1960s. *Crime Prevention and Community Safety*, *5*(3), 27–37.

Williams, R. (1974/1990). *Television: Technology and cultural form*. London: Routledge.

Willim, R. (2017). Imperfect imaginaries: Digitisation, mundanisation, and the ungraspable. In G. Koch (Ed.), *Digitisation: Theories and concepts for empirical cultural research* (pp. 53–77). London: Routledge.

World Prison Brief. (2017). World Prison Brief data. https://www.prisonstudies.org

Wrenby, K. (1955). *50 år med Stockholms tidningsbud: En liten krönika* [50 years of the Stockholm newspaper messenger: A little chronicle]. Stockholm.

Yousman, B. (2009). *Prime time prisons on US TV: Representation of incarceration*. New York: Peter Lang.

Zuboff, S. (2019). *The age of surveillance capitalism: The fight for a human future at the new frontier of power*. London: Profile Books.

INDEX

Note: page numbers in italics indicate figures.